The Mathematical Reality

Why Space and Time are an Illusion

Alexander Unzicker

Part I: A Brief History of Physics

Copyright © 2020 Alexander Unzicker
All rights reserved.
ISBN-13: 9798602252484

Table of Contents

Preface 3

Part I: A Brief History of Physics 7

1 Gods, Constants of Nature and other Defeats of the Mind 9

2 Large and Small Scale Simplicity: Gravity and the Quantum 23

3 Heat, Radiation and Matter: Modern Physics is Emerging 41

4 Cosmology Explains the Gravitational Constant 59

Part II: The End of Space and Time 71

5 The Cosmos Without Expansion: A World of Variable Scales 73

6 Revolutions That Have Not Yet Taken Place 85

7 The Origin of Mass and the Riddle of Physical Units ... 99

8 Finite Speed of Light: The Subtle Anomaly 111

9 Shrewd Atoms: Another Problem for Newton 121

Part III: The Mathematical Universe 135

10 Possible Alternatives for Space and Time 137

11 The Three-dimensional Unit Sphere – Full of Surprises 155

12 How S^3 Manifests in Reality 179

13 Unsolved and Crazy Stuff and Pure Mathematics . 199

Outlook 209

Acknowledgement 211

Literature 212

Picture Credits 214

Part I: A Brief History of Physics

Preface

This book is about fundamental physics. It aspires to form a consistent picture of reality by observing nature from the cosmos to elementary particles. The new approach I present here is based on investigating constants of nature and questioning their origin. A thorough analysis of the history of physics leads to the conclusion that there is a serious problem with what have been considered the basis of reality for centuries: Space and time. These may be the most accessible concepts for human perception, but are probably unsuitable for a basic understanding of nature.

From this analysis it also follows that current ideas in physics, especially the standard models of particle physics and cosmology, offer very little help for a real understanding. Skeptics of these models will find arguments supporting that claim, but a critique of contemporary physics is not my focus here.[1] I would rather like to point to the importance of constants of nature, explain what has been achieved so far in eliminating them, and then discuss the pending unsolved problems. Finally, I would like to present a mathematical alternative that could possibly replace space and time.

This is not a promise of an all-encompassing theory. Yet a new perspective unfolds that clarifies which problems of fundamental physics can and must be solved in order to achieve a satisfactory understanding of reality. Ultimately, we search for mathematical objects whose properties describe the various physical phenomena in purely mathematical terms. Because of the

---

[1] It can be found in detail in my books *Bankrupting Physics* (2013) and *The Higgs Fake* (2013).

Part I: A Brief History of Physics

enormous difficulties involved, such ideas are necessarily speculative, yet strictly avoid arbitrary assumptions that have no place in a rational description of reality.

Consequently, this book is also aimed specifically at mathematicians. Although their activities are often misguided by current theoretical fashions, they nevertheless have a crucial contribution to make to the understanding of nature, especially by studying the three-dimensional unit sphere that plays an essential role in these considerations.

Physicists who work in the tradition of natural philosophy by reflecting upon 'what holds the world together in its inmost folds' – as Einstein, Schrödinger, and Dirac did – will find a guidebook on what physics can achieve. But even non-scientists, once they become familiar the historical-methodological approach outlined in the following chapters, will easily understand that physics needs a new paradigm that goes beyond the concepts of space and time. Little prior technical knowledge is required, and the grey underlaid boxes which require mathematical background can be skipped without losing the overall context.

On the other hand, a little patience is required of scientists when they are invited to contemplate the history of physics from a new methodological point of view. Those interested in the groundbreaking consequences for space and time may even be tempted to skip the first or second part. However, such a historical analysis is key to an understanding of how the alleged entanglement of space and time emerged, and makes one realize the extent to which the current models deviate from a rational description of reality.

History is the only solid foundation on which such a new hypothesis about space and time can be built, which otherwise may sound outlandish. Therefore it is all the more important to estab-

Preface

lish a basis of indisputable facts relating to how scientific revolutions have occurred in the past. Hopefully, the following account will enable you to recognize a typical pattern behind these great achievements of humankind. To get an even clearer picture, it will also be helpful to have a look at the cognitive mechanisms with which the species *Homo sapiens* has struggled so far to fathom the laws of nature.

Munich, January 2020   *Alexander Unzicker*

Part I: A Brief History of Physics

Preface

Part I: A Brief History of Physics

"I cannot imagine a reasonable unified theory containing an explicit number which the whim of the Creator could just as easily have chosen differently." – Albert Einstein

Part I: A Brief History of Physics

1 Gods, Constants of Nature, and other Defeats of the Mind

Imagine looking at the starry sky in the Stone Age, without smog, urban light pollution, and all other annoyances of modern civilization. There can be no doubt that human beings back then were intrigued by the stars and tried to comprehend the laws of the spectacle going on in the skies. In the early hunter-gatherer societies, observing natural phenomena led people to create the first mythologies. Already in ancient Egypt, the appearance of Sirius, which usually preceded the flooding of the Nile, was understood as a signal to start cultivating the fields. This is how we react as human beings: a continuous process of reasoning takes place in our mind, and we try to make sense of what we perceive, connecting facts that may or may not be causally related after all. Back then, it seemed more than obvious that what was going on in the skies was ruled by higher powers such as Re, the solar deity. Who other than a mighty deity could steer the course of the inaccessible stars?

Later, the ancient Greeks began to identify the *wanderers* in the starry sky, called planets, and ascribed divine qualities to them. These systematic observations were nothing other than an early form of scientific research. However, at any period in history, science has had its respective limits of comprehension, beyond which lay unknown territory. In the ancient cultures, some phenomena were called gods because they could not be explained otherwise. Yet, the parallels to modern science are quite obvious. The assumption that it was not just the whim of individual gods that governed planetary motion but rather profound laws may be seen as an early attempt to create a "unified" theory of the universe — something that physicists dream of to this day.

Part I: A Brief History of Physics

MEDIEVAL ASTRONOMY AND ITS ARBITRARY NUMBERS

As time went by, people paid more and more attention to planetary motions and recorded them accurately. Yet, despite the idea of a single, almighty God, astronomy was forced to attribute a series of characteristics to the individual planets until the Middle Ages. All orbits were assumed to be precise circles around the Earth, but in order to account for the apparent retrograde motion[1] of some planets (that indeed hinted at the Sun being at the center of their orbits), additional assumptions for their orbits were invoked: so-called epicycles, that is, smaller circles mounted on their larger relatives. As observations became still more accurate, the above description turned out to be insufficient again, prompting the postulation of other workarounds such as the eccentric, a quantity that denoted how far the circle on top of the circle had shifted its original center. Of course, all these assumptions were far from intellectually satisfying, but for lack of any better explanation, astronomers reluctantly accepted the existence of such arbitrary quantities — God-made numbers beyond the limit of human understanding.

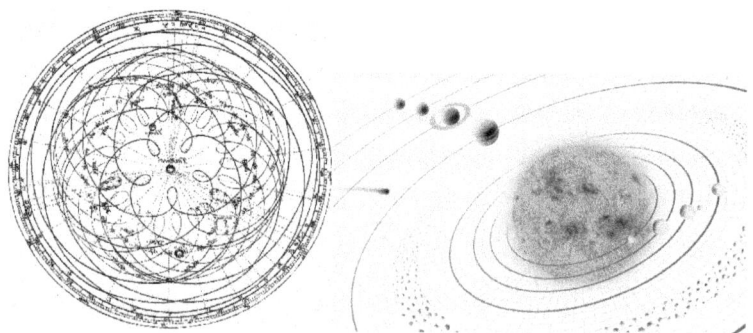

Geocentric (left) and heliocentric (right) world view after Copernicus. The extremely complicated planetary orbits in the geocentric model were suspicious to many researchers.

1 Gods, Constants of Nature, and other Defeats of the Mind

In his monograph *Big Bang*, Simon Singh comments on the epicycles as follows:[2] "Every flawed model can be saved by such fiddling around with numbers." However, we shall not be tempted to disregard these early forms of science — after all, the geocentric world view back then was by no means stupid or even far-fetched. Observing the sky was already a great achievement in itself. It has to be complemented by looking for deeper theoretical reasons, yet this search is not always in lockstep with observational progress.

The geocentric picture that dominated astronomy until the Middle Ages was retrospectively seen as a deadlock, because instead of providing explanations, the model indulged in a description by arbitrary numbers — God-given parameters with which the Almighty seemingly had endowed the planets and their orbits. This increasing complexity, so the anecdote goes, was once commented on in the following terms by King Alfonso el Sabio of Castile: "If the Lord Almighty had consulted me before embarking creation thus, I should have recommended something simpler."

MATHEMATICS OPENS NEW HORIZONS

On the other hand, not to overly question the limits of current knowledge seems to be deeply rooted in the nature of *Homo sapiens*: Gods, God-given numbers — whatever cannot be explained here and now — are readily declared to be part of the unfathomable. Note that it was the scientific elite who used to postulate such absolute limits of knowledge, unconsciously perhaps, because conceding that one's wisdom is insufficient is something uncomfortable to face. Consequently, recognizing the merits of the Ptolemaic world view is necessary for a proper historical apprehension of science, though a dramatic leap of understanding occurred when the Copernican revolution unfolded.

Part I: A Brief History of Physics

Copernicus intuitively understood that the Sun, not the Earth, was the center of planetary orbits, a point of view that immediately simplified the maddeningly complicated picture of motions. Based on Danish astronomer Tycho Brahe's precise measurements, Johannes Kepler eventually realized that the orbits around the Sun were ellipses rather than circles — a spectacular insight that all of a sudden revealed the essential role of mathematics in the laws of nature. "The book of nature is written in mathematical language", as Galileo Galilei famously phrased it, after having contributed to the breakthrough of the heliocentric model with his newly developed telescope.

It was then Isaac Newton's turn to complete the revolution with an elaborately constructed system of both mathematical methods and physical concepts that led to an understanding of planetary motion on a highly advanced intellectual level. Inspired by the visionary idea of celestial and earthly motion having the same origin, he was able to prove that the planets followed the course of Kepler's elliptical orbits exactly — a triumph of the human mind that is certainly unparalleled to this day and marked the beginning of modern natural science.

CONVERGING TO SIMPLICITY

Let us recall the crucial elements of this scientific revolution. Frequently mentioned, yet a minor point, is that Kepler's ellipses described the orbits more precisely than the old Ptolemaic system. More importantly, the prominent role of mathematics in the laws of nature had become evident and led to a unification of earthly and celestial gravity, an unprecedented insight that deeply satisfied the human desire for understanding. The long-lasting search for causes that presumably had started with primitive hypotheses in the Stone Age had culminated in a beautiful way. Methodologically, however, the decisive aspect — and this argu-

1 Gods, Constants of Nature, and other Defeats of the Mind

ment will guide you through the entire book — is that revolutionary breakthroughs always go along with a *simplification* of the theory. That is why in Newton's theory, fewer arbitrary assumptions and unexplained parameters are needed – indeed, just one, namely the gravitational constant G named after him ('Big G'). The latest measurements determine it as $6.673 \cdot 10^{-11}$ m^3/(s^2 kg).[1]

Three hundred years later, in his treatise *The Structure of Scientific Revolutions*, philosopher Thomas Kuhn brilliantly showed the interplay of surprising new data (anomalies) and the increasingly complicated models invoked to describe them. The more sophistication and the more ad-hoc assumptions such models use, the more unstable they become, eventually collapsing into something much simpler, which is intuitively recognized as the better model. In the old geocentric system, ignorance was disguised by dozens of seemingly God-given numbers. During the revolution, they became obsolete and were replaced by a single parameter: the gravitational constant G. The epistemological progress is close at hand: dramatically fewer arbitrary assumptions had to be made about nature. Newton had, so to speak, sent a good many gods or God-given numbers into retirement and replaced them with a "monotheistic" concept (the gravitational constant) — as we know from history, to the utter displeasure of the real clergy. Henceforth, cardinals and archbishops lost their status as intellectual leaders of humankind and had to cede it to those whom we call scientists today.

---

[1] Not to be confused with the local gravity g=9.81 m/s^2, which is only significant on our planet and can be calculated from the radius of the Earth r, its mass M and G: g= GM/r^2. G thus has a far more fundamental meaning than g. The fact that G was measured much later is not significant for this methodological consideration.

Part I: A Brief History of Physics

CONSTANTS OF NATURE - MOMENTARY LIMITS OF KNOWLEDGE

The history of science shows that it is not the names that matter, but the function in the system. Scientists today act as the world's enlighteners, just like the medieval theologians; what used to be called God, we now call laws of nature. The numbers that show up in these laws today are called constants of nature or even fundamental constants, but epistemologically, they differ very little from the God-given parameters of the epicycle model. There are still limits to our knowledge.

Modern physics formulates its laws using a variety of constants of nature that are not justified by deeper reasons. Since they cannot be explained by our scientific elite, they are considered inexplicable. Needless to say, these parameters represent a much higher level of knowledge than the many "planetary gods," simply because there are fewer of them, not to mention the degree of mathematical abstraction required to distill them from observation. But, at the end of the day, the transition from gods to fundamental constants represents only gradual progress. We believe in constants of nature because they have resisted all our efforts to explain them: we neither understand their numerical values nor do we know the reason for their sheer existence. Their enigmatic connection to the universe gives them an almost mystical meaning, but let us be clear: constants of nature are the gods of modernity.

Although Newton's great success reduced the number of unexplained constants to one (G), further constants emerged as the natural sciences thrived. In 1676, after meticulously observing the moons of Jupiter, Danish astronomer Ole Rømer concluded that light propagated at finite speed, a fact that Newton accepted but had not expected. As a consequence of the Copernican revolution (heavenly forces could be investigated in mundane labs!),

1 Gods, Constants of Nature, and other Defeats of the Mind

researchers were eager to ask nature questions by experimenting. In the 18th century, electricity and magnetism were discovered, and despite the elegance of the laws that described them, new constants of nature were needed in their formulation. In the early 1900s, researchers stumbled upon a couple of intriguing phenomena in atomic physics. It was Einstein who first recognized the significance of a new constant of nature introduced by Max Planck: the so-called quantum of action h, widely believed to be one of the most important constants.

All these developments were accompanied by great insights, profound mathematical theorems, and sometimes a wonderful unification of seemingly unrelated phenomena. In the following chapters, we will look more closely at how these insights came about with fewer and fewer parameters, or 'constants of nature'. Despite all progress, however, physics still needs some fundamental constants to describe nature's behavior. And it is precisely here that our knowledge, which is certainly far advanced, reaches its limits.

Unlike most physicists, I am convinced that these constants of nature do not represent an absolute limit to our knowledge, but mark our currently still limited understanding. Ultimately, these constants of nature are arbitrary, unexplained numbers that have allowed academics to find peace of mind by declaring the unexplained to be unexplainable. However, a thorough historical and methodological reflection forces us to consider an alternative: The alleged existence of fundamental constants simply means that we have not yet understood the laws of nature down to their origin. There are no constants of nature, just as there are no gods.

PHYSICS MEANS NATURAL PHILOSOPHY

Before we delve into the consequences of this proposition, keep in mind that it is not a scientific result in the conventional sense. Epistemologically, such constants of nature could of course exist, just as there *could* be Gods who rule the world. Assuming there are no fundamental constants means taking a stance that originates from natural philosophy, a working hypothesis under which I should like to scrutinize the current state of physics. I think that this hypothesis is backed by plenty of methodological arguments and historical evidence, yet it is by no means universally accepted — as was the case with atheistic world views.

As a matter of fact, the existence of fundamental constants as such is hardly challenged by any of today's theoretical physicists. However, I can assure you that the idea of questioning fundamental constants has prominent supporters — not least Albert Einstein, who pondered these problems intensely in his later years.[3] His views about constants of nature went unappreciated for no good reason, as was the case with the philosophical approach to physics he shared with other great thinkers of his time: Erwin Schrödinger, Paul Dirac, and, in particular, the Viennese philosopher Ernst Mach, who contributed to the understanding of gravity in an extraordinary way. It may surprise you how much the chapters that follow focus on the early 20th century and ignore the later developments in physics. However, the latter are largely irrelevant with regard to fundamental questions. Please bear in mind that my argument is ultimately based on this core proposition: a rational description of nature cannot tolerate fundamental constants. Gods – whatever we call them – have no place in reality.

A SIMPLE DEFINITION OF SIMPLICITY

Therefore, the fewer constants of nature a physical theory requires, the more confidence it can inspire. Simplicity is the quality criterion of theories and it can be defined quite easily by counting free parameters. The fewer arbitrary numbers a theory uses, the simpler it is. There is no difference in principle between

1 Gods, Constants of Nature, and other Defeats of the Mind

constants of nature and free parameters, which are often introduced ad hoc. The former can often look back at a long history of discoveries and are considered, with some justification, more fundamental. Sometimes they have a wide range of applications, but they are nevertheless unexplained numbers. In the following, we will therefore regard every quantified observation as a constant of nature. The number of such free parameters can be easily determined, and even physicists with completely different world views will be able to agree on how to count them. At the moment, the standard models of particle physics and cosmology have a total of more than a hundred free parameters, and you might think that the criterion of simplicity, which is obviously lacking here, comes in handy for me to attack these constructions.

However, it is not a philosophical whim of mine to postulate simplicity, but rather the only consistent interpretation of the history of science. There is historical evidence that revolutionary insights have always been accompanied by simplification in the sense described above, i.e. reducing the number of free parameters. This process of simplification is a guiding thread throughout all fields of physics (probably science in general), and we will examine it in detail in the following chapters.

DISTRACTIONS OF CONTEMPORARY PHYSICS

As a necessary interlude, we need to address some of the commonplaces in modern physics that obscure the real problem, though they might seem to agree with the arguments outlined above. Since we use the number of constants of nature as a criterion for simplicity, it is essential to clarify what a constant of nature is.

A widely propagated, though serious misconception in current theoretical physics is to regard only 'dimensionless' constants of nature as 'fundamental', which means only those that have purely

Part I: A Brief History of Physics

numerical values. Prominent examples are the proton-to-electron mass ratio of 1836.152673... (two particles that form the hydrogen atom), or the notorious fine structure[I] constant 137.035999... that Richard Feynman had called a "great damn mystery" of physics. These are indeed tantalizing numbers.[II]

However, it is completely misleading to claim that[4] the gravitational constant G, amounting to $6.673 \cdot 10^{-11}$ m³/(kg s²) is not fundamental because it contains the physical units meter, second and kilogram. Evidently, the numerical value 6.673 has no particular meaning, because numbers would be different if meters, seconds, and kilograms had been defined differently. But G remains, of course, a fundamental message of nature about the strength of gravity, which is exactly as big as we measure it, and not half, double, or ten times that value. Instead of glossing over this unanswered question, the concrete value of the gravitational constant G ought to be explained if we ever want to achieve a thorough understanding of gravity.

UNDERSTANDING INSTEAD OF EXCUSES

Occasionally it is argued that there are an infinite number of parallel universes in which all numerical values of the constants of nature are realized, and that we live in the one with 137.035999. According to an 'anthropic principle', our universe is the only one that is not 'hostile to life' because – lucky as we are supposed to be – only the numerical values here were suited for the emergence of intelligent life. Not only is there not the slightest evidence for such fantasies,[III] but they also convey poor

---

[I] Technically speaking, the inverse of the fine structure constant $\alpha = 1/137$.
[II] The following anecdote is reported about Dirac: When a young physicist presented him with an idea for a new theory, the scholar was immediately interrupted by Dirac: "Can you calculate the fine structure constant? No? Well, come back when you've done it!"
[III] Presumably, everybody would be happy in a world with a smaller G, except

1 Gods, Constants of Nature, and other Defeats of the Mind

logic: it is characteristic of intelligent beings to seek explanations for arbitrary numbers.

However, with so much science fiction penetrating physics it is easy to overlook the missing real explanations. A prominent example is the speed of light c, also a fundamental constant. For the sake of convenience, it has now been *defined* as a numerical value, 299792458 m/s, and the actual measurement of c has been shifted to the definitions of meter and second. Yet the fact that nature dictates a certain speed of light propagation is a surprising and deeply mysterious property of the universe. Can we deduce it from pure logic? Of course not, otherwise Newton might already have done so. Today, therefore, c is a limit of our knowledge, and we must explore the deeper reason behind its existence if we want to understand the universe properly.

The same holds true for Planck's quantum of action h, a constant we will discuss later in detail. Again, its physical units are all but irrelevant. As we shall see in the following chapters, considering physical units was often the key to decisive breakthroughs. Nonetheless, today's theoreticians are used to making 'unit-free' calculations by setting various constants of nature equal to unity. From a natural philosopher's point of view, this is a nonsensical approach that obscures fundamental questions and stalls any progress in solving them.

BACK BEFORE NEWTON?

Once we reflect about the origin of the speed of light and question its status as an untouchable constant of nature, we face serious consequences. Two concepts have been the basis of all physical reality for centuries: Space and time, whose units of meters and seconds are intertwined by the speed of light. But there is a

---

orthopedists and bone surgeons.

Part I: A Brief History of Physics

fundamental need to explain *why* reality presents itself in such a 3+1-dimensional fashion. Even more mysterious than the number of dimensions is the *qualitative* difference between space and time.

Newton was unable to derive these elementary notions from first principles and postulated a Euclidean (flat) space and a uniformly running time as basic quantities that nobody was supposed to question. However, it may well be that this very basis of his great discoveries contained the seeds of failure. Space and time are without doubt notions that are easily accessible to human perception, but couldn't there be more elementary quantities hiding behind them that represent reality by providing deeper insights? In fact, there are many clues that space and time are fundamentally unsuitable concepts to decipher the Book of Nature correctly.

As a matter of fact, the speed of light c is an unexplained, ultimately arbitrary parameter. In history, such numbers have always appeared when an underlying assumption was wrong. Back in the Middle Ages, the erroneous view that the Sun rotates around the Earth produced a large number of parameters. But even a single superfluous parameter like c points to a possible flaw in the underlying assumption: specifically the concepts of space and time.

Thus, the existence of the speed of light is a contradiction of Newtonian mechanics, all the more because it is also a limiting velocity for matter. There is no reason in Newtonian mechanics that would prohibit accelerating objects to arbitrarily high speeds. The fact that this is impossible is a textbook example of an anomaly in the sense of Thomas Kuhn, indicating that Newtonian physics is incorrect. Einstein saved Newton's work from being bluntly falsified by describing the laws of dynamics with his special theory of relativity. But this came at a price of attributing an arbitrary property to nature: the speed of light c.

1 Gods, Constants of Nature, and other Defeats of the Mind

DISSOLUTION OF CLASSICAL PHYSICS

What happened to physics on the large scale also loomed at the microscopic level around 1900: Newtonian physics no longer worked, and all of a sudden a new constant of nature appeared once again: Planck's quantum of action with the value $h=6.626 \cdot 10^{-34}$ kg m^2/s. Despite its name, it is again Albert Einstein who deserves the credit for having recognized the importance of the quantum of action. A whole new theory, known since 1920 as quantum mechanics, was built on the foundation stone h. Quantum theory is believed to be incompatible with the theory of relativity, which is based on the constant c. Conventionally, this incompatibility is perceived as the largest obstacle physics has to overcome before arriving at a unified theory. However, this is probably thinking in the wrong direction. It is more likely that these two most important theories of physics cannot be unified at all, because they are based on false premises: the concepts of space and time. Both h and c would then be symptoms of the deficiencies of Newtonian physics.

However, before we go into more detail on the intricacies of space and time and the possible alternatives, it is wise to undertake a thorough examination of history by focusing on constants of nature. Again, the central thesis of this book applies: if we seek a rational description of reality, constants of nature must ultimately be explained.

Part I: A Brief History of Physics

2 Large and Small Scale Simplicity: Gravity and the Quantum

The following considerations about the history of physics are intended as a systematic approach. It is not intended as a comprehensive account, nor is there a need to proceed in chronological order. I shall rather focus on scientific revolutions and how they came about. It turns out that such revolutions usually contain three elements: a visionary idea, a mathematical formulation, and ultimately, an elimination of arbitrary constants, which often leads to a unification of theories.

The smaller number of free parameters after the revolution is exactly what we need: a technical, yet easily understandable definition of what simplification of a theory means. It is such a simplification that goes hand in hand with real progress, yet the above three-step pattern may appear in different facets. Sometimes, arbitrary numbers are eliminated as soon as they are discovered, when the theory is already established. On the other hand, there might be huge delays within the idea-mathematization-simplification scheme: partial revolutions can take place that need centuries to be completed. Take, for example, the visionary idea of the Greek philosopher Democritus, according to which nature consisted of indivisible small building blocks called atoms: this could only be considered complete by 1930, with the development of modern atomic theory.

The textbook example of a scientific breakthrough, and at the same time the onset of modern science, is Isaac Newton's classical mechanics and law of gravitation. In Part III we will return to this crucial moment and reflect on possible further progress in fundamental physics.

Part I: A Brief History of Physics

IT STARTS WITH AN IDEA

The visionary idea and perhaps the real stroke of genius in Newton's work was suspecting that mundane and celestial motions obeyed the same laws. Certainly, the development of classical mechanics, with its concepts of mass, force, velocity and acceleration, was in itself a tremendous achievement and prerequisite for what followed. However, linking together the testable, earthly laws of motion with celestial mechanics that, until then, had belonged to the realm of the supernatural, marks the revolution with the greatest impact on the intellectual history of mankind in general.

Since the parallel between the attraction on the moon and the falling apple can be explained to a child, it is easy to overlook the courage that was necessary to develop such an idea in the 17th century. Thus, the visionary idea is often the underappreciated element of a scientific revolution that requires extraordinary creativity and boldness. In hindsight, being ahead of one's time looks easy.

MATHEMATICS IS NECESSARY

Admittedly, even a good visionary idea cannot prevail without mathematical formalism, which Newton also succeeded in developing, although the English polymath Robert Hooke had provided him the key idea. Presumably, it was not too hard to guess that the gravitational pull of the Earth would become weaker with increasing distance, but to confirm this fascinating fact, some nontrivial math was needed. Newton calculated the Moon's centripetal acceleration from its known distance from and orbital period around the Earth. He saw that the result was about 3600 times smaller than the local gravity of $g=9.81 \, m/s^2$ that caused the proverbial apple to fall from the tree on the Earth's surface. Since the distance between the Earth and the

2 Large and Small Scale Simplicity: Gravity and the Quantum

Moon amounts to about 60 Earth radii, after noting that 60^2=3600, it became obvious that the decrease in the Earth's gravitational pull could not be proportional to the distance, but rather to its square.

> A. 5. Hooke to Newton, Jan. 6, 1680.
>
> SIR,
> Your calculation of the curve described by a body attracted by an aequall power at all distances from the center, such as that of a ball rolling in an inverted concave cone, is right, and the two auges will not unite by about a third of a revolution; but my supposition is that the attraction always is in duplicate proportion to the distance from the center reciprocall, and consequently that the velocity will be in a subduplicate [proportion] to the attraction, and consequently as Kepler supposes reciprocall to the distance: and that with such an attraction the auges will unite in the same part of the circle, and that the nearest point of the access to the center will be opposite to the furthest distant, which I conceive doth very intelligibly and truly make out all the appearances of the heavens. And therefore (though in truth I

Excerpt from a letter from Hooke to Newton. Hooke was the first to mention the inverse-square law of gravity and also associated it with Kepler's findings. Later, Newton refused to give Hooke any credit for this essential contribution.

This so-called inverse-square law is expressed in the formula

$$F_g = \frac{GMm}{r^2}$$

where M is the mass of the Earth, m is the mass of a body[1] attracted by it, r is the distance from the Earth's center, and G the gravitational constant, which we will discuss in much detail later. The formula was proven to be correct for all celestial bodies in the solar system, but also for the local gravity that everybody can feel. This fascinating coincidence is incorporated in the formula

$$m \cdot g = \frac{GMm}{r_E^2} \qquad \text{or simply} \qquad g = \frac{GM}{r_E^2},$$

---

[1] The symmetry of the two masses M and m is due to the "inherent property of all bodies to attract each other", a quote often attributed to Newton, although no concrete source can be found.

Part I: A Brief History of Physics

r_E being the Earth's radius.

CONSTANT OF NATURE OR NOT?

Now, let us determine which of these quantities we should call arbitrary parameters or 'constants of nature'. The Earth's radius r_E and its mass M *do not* fall into this category, because they are measurement values that obviously originated from random events while our planetary system was being formed. The question "why exactly this number and neither more nor less?" does not arise.

This question needs to be raised for the value of local surface gravity $g=9.81 m/s^2$, as well as for the general gravitational constant $G=6.673 \cdot 10^{-11}$ $m^3/(s^2 kg)$. In principle, both can be seen as constants of nature or unexplained, free parameters. The crucial insight provided by Newton's law of gravity is that g is now explained by the above formula – i.e. it can be calculated and eliminated from the territory of the unknown. The number of constants of nature has been reduced by one. In fact, today no one would call g a fundamental constant, but rather the surface acceleration on a particular planet. Here is the epistemological progress of Newton's law of gravity: it reduces the number of arbitrary parameters and thus simplifies the theory.

Let us illustrate this with another thought experiment. If we forget the law of gravity for a moment and measure the surface acceleration on various celestial bodies such as the Earth's moon, or on Mars, Jupiter and Saturn. All these numerical values for local gravity would count as 'constants of nature' that await an explanation – as long as one is unaware of Newton's law that makes them obsolete. It becomes clear that the law of gravity provides a tremendous simplification.

2 Large and Small Scale Simplicity: Gravity and the Quantum

When contemplating the big picture vision-mathematization-simplification of Newton's law, it is irrelevant that the gravitational constant G=6.673·10^{-11} m^3/s^2kg was actually measured more than a century later, in 1798, by Henry Cavendish. Indeed, Newton had not believed that the gravitational force, which we can well feel from a body of the size of the Earth, would ever be measurable for everyday objects. Cavendish eventually succeeded in determining G by measuring the force between two known masses using a sophisticated device called a torsion balance.[I] Then, with the product GM known from astronomy, the mass of the Earth M could be calculated for the first time. Of course, weighing the Earth and other celestial bodies by direct methods is impossible to this day.

REVOLUTION BY THE COURT MATHEMATICIAN IN PRAGUE

For the solar mass M_s, the product GM_s is called Kepler's constant, and here again it is worthwhile to see how much Newton's predecessors had prepared the ground with their astronomical findings. The vision-mathematization-simplification pattern can be observed in Johannes Kepler's work, too, which is hardly less revolutionary than that of Newton. Prior to that, however, one must consider the geocentric world view of the Middle Ages, which described the planetary orbits quite precisely, yet with a multitude of arbitrary parameters – today we would call them constants of nature.[II] The visionary idea here was Copernicus' insight that the celestial bodies do not revolve around the Earth, but rather that all planets, including the Earth, orbit the Sun. This might be regarded as a 'simpler' model already, yet it does not

---

[I] The basic idea goes back to English physicist John Michell (1724-1793).
[II] Julian Barbour, in his book *The Discovery of Dynamics,* describes the historical formation of terms such as 'deferent', 'equant', 'epicycle', etc.

Part I: A Brief History of Physics

apply in the strict sense of our definition (fewer constants of nature). One reason why the Copernican idea did not immediately win the day was its inability to describe the planetary orbits precisely. This became possible only when Kepler eventually freed his mind from the dogma of circular orbits and considered ellipses. In this revolution, the element of mathematization played a key role and required exceptional knowledge of geometry.

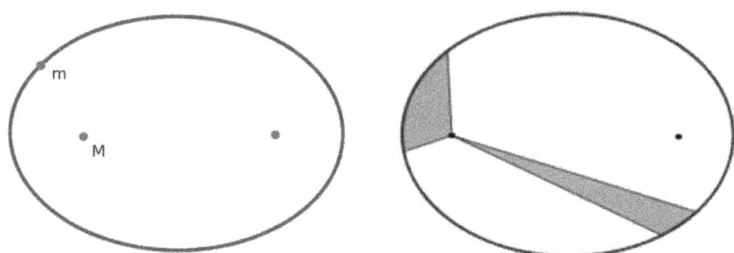

Kepler's laws of planetary motion: Planets move on ellipses with the Sun at their focal point (I). A pointer to the planet sweeps out equal areas in equal times (II).

Kepler's first law, which states that planetary orbits are ellipses with the Sun at one of the two foci, is an obvious simplification: the shape of an ellipse is determined by two parameters (e.g. semimajor axis and eccentricity), whereas in the Ptolemaic model up to four numbers were needed for each planet. If we count bits of information, this is simpler, because fewer arbitrary numbers are needed to describe natural phenomena – and herein lies the scientific breakthrough. However, the new description had to be completed by Kepler's second law, which assigned a certain speed to each planetary position, demanding that the line from the Sun to the planet must sweep out equal areas in equal times.

One could now argue that besides the two parameters of the ellipse, the orbital period of the planet is another unexplained number. However, after ten years of pondering over the math, Kepler resolved this problem with his third law: the outer planets

2 Large and Small Scale Simplicity: Gravity and the Quantum

move more slowly, the squares of the orbital period T being proportional to the cubes of the semimajor axes a. Thus $\frac{a^3}{T^2} = K$ holds for *all* planets in the solar system, which is a dramatic simplification. One single constant describes the motion of all planets! However, the spectacle in the skies is still more complicated: take for instance the moons of Jupiter, first observed by Galilei in 1610. If one considers their motion around Jupiter,[1] an analogous relation $\frac{a^3}{T^2} = K_J$ applies to all moons, but with a different constant K_J. In order to understand that, it took Newton to identify the constant K with the product $\frac{GM}{4\pi^2}$. Even with Kepler's findings, the constants K of the various planets would still be arbitrary parameters. Only Newton's law of gravity explains *why* the different planets let their respective moons orbit at different speeds: it is due to the gravitational pull of their mass! The fact that all dynamically determined constants K can be attributed to a sole gravitational constant G represents the same simplification we noted above with the different surface accelerations g.

It should be not be forgotten that Newton's main mathematical achievement (technically, it was actually the most difficult task) was deriving planetary motion from his law of forces, thus explaining the parameters of Kepler's ellipses in physical terms (energy and angular momentum).

In order to properly appreciate the scientific value of the Copernican revolution completed by Newton, we must not only count the parameters known back then, but also consider how many measuring values all known celestial bodies *today* would produce – all of which would have to be regarded as 'constants of nature' if Newton's mechanics were not known. Today's methods of observation have revealed literally millions of them!

---

[1] Being at the focus of the respective ellipses of the moons.

Part I: A Brief History of Physics

Savor the fact that this huge amount of data can be described with only one constant G. In terms of information theory, this is a colossal simplification and, of course, a triumph of the human mind that marked the beginning of modern civilization.

THE NUMBER CRUNCHER

Now we leap ahead more than two hundred years from Newton's time, from the vastness of the solar system to the microcosm of the atom, where spectacular progress was made at the beginning of the 20th century. Again, the pattern of vision-mathematization-simplification will be recognized. In addition, it is here that the great insights about the solar system find their fascinating microscopic counterpart. The idea that electrons orbit the atomic nucleus, just like planets orbit the Sun, evoked an enormous fascination among physicists at that time. While the first idea goes back to Wilhelm Weber in the 19th century,[5] around 1904 the Japanese physicist Hakamura independently came up with the model, and Niels Bohr integrated it into a grand new picture of the atom. Even if that model turned out to be incomplete in parts, it certainly constituted a visionary idea during the scientific revolution in atomic theory.

Just as Newton relied on Kepler, Bohr's model of the atom had an important precursor – the Swiss math teacher Johann Jakob Balmer, without whom this breakthrough would not have been made. Like Kepler, Balmer closely examined the observations and spent years agonizing over hidden mathematical patterns in the data. With an incredible instinct, he revealed a mysterious relation, though he could not yet explain its true origin. In 1885, Balmer studied the spectrum of the hydrogen atom, i.e. he observed the different wavelengths of light emitted by hydrogen gas under electric discharge. Although the spectral colors were beautifully bright (the wavelength determines the color of the light), they appeared in a seemingly

2 Large and Small Scale Simplicity: Gravity and the Quantum

random sequence of 656, 486, 434, and 410 nanometers. According to our technical definition, these values must be considered constants of nature, but physicists at the time showed surprisingly little interest in these numbers.[1] Balmer realized that these numbers were important messages from nature and began to search for regularities. He probably tried to put them into proportion and searched for fractions with small integers, whereby coincidences such as $\frac{486,1}{656,3} \approx \frac{20}{27}$ or $\frac{434,0}{656,3} \approx \frac{125}{189}$ may have emerged. What exactly was going on in his brain will probably remain a mystery forever. But one can imagine how, without any technical gadgets at hand, it took him years to arrive at the following amazing formulas by trial and error:

$$\frac{20}{27} = \frac{\frac{1}{2^2}-\frac{1}{3^2}}{\frac{1}{2^2}-\frac{1}{4^2}} \text{ and } \frac{125}{189} = \frac{\frac{1}{2^2}-\frac{1}{3^2}}{\frac{1}{2^2}-\frac{1}{5^2}}.$$

This stunning result enabled him to calculate the wavelengths of the spectral lines as follows:

$$\frac{1}{656{,}3 \text{ nm}} = R(\frac{1}{2^2}-\frac{1}{3^2}), \frac{1}{486{,}1 \text{ nm}} = R(\frac{1}{2^2}-\frac{1}{4^2}),$$

$$\frac{1}{434{,}0 \text{ nm}} = R(\frac{1}{2^2}-\frac{1}{5^2}), \frac{1}{410{,}2 \text{ nm}} = R(\frac{1}{2^2}-\frac{1}{6^2}),$$

while introducing the constant $R = 1.0973731 \cdot 10^7 \text{m}^{-1}$ that was later named after the Swedish physicist Rydberg.[II] Imagine what kind of satisfaction Balmer must have felt when, after such an exhausting search, he discovered this fantastic relation!

Two more things remain to be noted here. First, the appearance of squares in the denominator actually reflects the similarity

---

[1] As the German Nobel Laureate Theodor Hänsch once jokingly remarked in a lecture, physicists considered these spectroscopic data as "something dirty, almost chemistry".

[II] To be precise, Balmer discovered the constant R_H valid for the hydrogen atom, which also accounts for the motion of the nucleus, as was discovered later. Therefore, a minute, though calculable difference between R_H and R (for heavier atoms) remains, which is irrelevant for our methodological arguments.

Part I: A Brief History of Physics

between Newton's law of gravitation and the corresponding electrical law of Coulomb, a resemblance that led to the idea of atoms as tiny solar systems. However, the way Balmer arrived at his conclusions is also remarkable from a methodological point of view. Like Kepler, Balmer looked out for mathematical coincidences without it being clear that they existed at all. Nowadays, such an approach is often dismissed as 'number cult' or 'numerology'. But Balmer's success, as in Kepler's case, relied precisely on that approach: finding a relation that was obviously correct, as further measurements of spectral lines would soon confirm. By doing so, Balmer dramatically simplified the description of nature's messages, providing a glimpse of the emerging mathematical framework. Having the courage to search for such mathematical patterns may in itself be called visionary.

Johann Jakob Balmer (1825-1898)

However, Balmer's discovery has undeniably reduced the number of constants of nature, which is our definition of simplification. All spectral lines of the hydrogen atom (in principle, this applies to all atoms) are described by Balmer's formula and its generalization;[I] the only remaining constant R is now called the Rydberg constant.

THE MYSTERIOUS NUMBER IN ALL ATOMS

In the history of atomic physics, as in astronomy, the vision-mathematization-simplification pattern has been repeated several times while it is simultaneously cross-connected with other areas of physics.

At the beginning of the 20th century, physicists discovered a particularly important constant of nature, the so-called quantum of action h. The mysteries involving h will keep us busy until the end of this book, and it is Albert Einstein who deserves the major credit for recognizing its relevance. In 1905, he boldly postulated that light could release only determined amounts of energy E, so-called quanta. The amount of energy depends on the frequency f or wavelength λ of the light,[II] as expressed in Einstein's famous formula $E=hf$. The constant h thus has the interesting physical unit energy (Nm) per frequency (1/s), thus Nms, which is also called *action*.

Considering the physical units of a quantity has proved to be the royal road to deep insights about the laws of nature – although

---

[I] In modern terms, Balmer's formula describes only the second atomic shell, because it is only this shell that emits light visible to the eye. With the discovery of ultraviolet spectral lines of the first shell (Lyman series) and infrared light from higher shell transitions (Paschen series etc.), the generalization of Balmer's finding to the general formula was obvious.
[II] The two quantities being related by the formula $c = \lambda f$.

Part I: A Brief History of Physics

this fact is not appreciated in the current scientific tradition. However, this is how Niels Bohr presumably launched the revolution in atomic physics. Due to the identity[I] N=kg m/s², the units of h may be rewritten as kg m²/s. Bohr noticed that this unit can be seen not only as energy per frequency (as in Einstein's formula), but also as the product of mass, distance, and velocity, forming the unit of angular momentum.[II] Like energy and linear momentum, angular momentum is an important conserved quantity in physics. It is effectively visualized by the onset of rapid rotation of ice skaters as soon as they pull their arms towards their bodies.

UNITS PROVOKING A STROKE OF GENIUS

Bohr now had the ingenious idea of relating h to the angular momentum L of an electron orbiting the atomic nucleus. This was the ingredient that filled the visionary idea of viewing atoms as small solar systems with life. However, the greatest puzzle he solved was to explain why the electron's motion around the nucleus was constrained to certain distances from it. While planets, in principle, can be found at any distance from the Sun (and this distance determines the orbital period in terms of Kepler's third law), the same does not apply in the case of electrons: Bohr realized that they can orbit the nucleus only if their angular momentum is a multiple of the constant $h/2\pi$, which is commonly abbreviated as $\hbar$ ("h bar"). Accordingly, the electrons' trajectories are characterized by L= $\hbar$, 2 $\hbar$, 3 $\hbar$ ... which also correspond to different energy levels.

[I] This follows from Newton's second law, F= m·a.
[II] For the sake of historical accuracy, the English mathematician John William Nicholson, who also considered angular momentum (Kumar, loc. 1904), must be mentioned. However, this does not diminish Bohr's accomplishment of having discovered a coherent picture of the atom.

2 Large and Small Scale Simplicity: Gravity and the Quantum

Although Bohr was unable to provide a deeper reason for this oddity, his model led to a spectacular discovery. When jumping from an outer to an inner shell, electrons had to release energy that corresponded to a frequency or wavelength according to Einstein's formula *E=hf*. Bohr showed that the possible jumps exactly matched the wavelengths of the light found by Johann Jakob Balmer! The numbers labelling the respective orbits, later called shells, turned out to be the squared integers that had appeared in the denominator of Balmer's formula. This was huge.

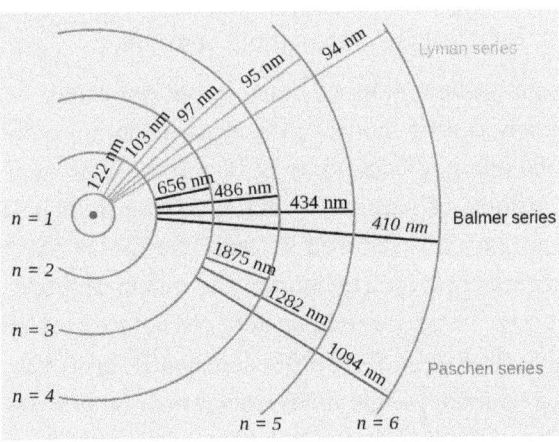

Schematic representation of the transitions in the hydrogen atom. Electrons are assumed to jump from a higher to a lower shell. The Balmer series, yielding visible light (about 400nm–800nm), was the first to be discovered.

The energy that is released during such an electron jump – say, from the fourth to the second shell – can simply be calculated as

$$E = hf = hc/\lambda = hcR\left(\frac{1}{4^2} - \frac{1}{2^2}\right),$$

which corresponds to the beautiful turquoise wavelength of 486.1 nm. Rydberg's constant R, which Balmer had measured but not explained, was no longer a big mystery, but could be derived from a short calculation. The analogue of Newton's law of gravitation needed for the atom is Coulomb's inverse-square law of electric force. Combining it with Bohr's postulate for the angular momentum yields the energy levels that can be compared with Balmer's formula, resulting in

$$R = \frac{m_e e^4}{8c\,\varepsilon_0^2 h^3},$$

where c = $3 \cdot 10^8$ m/s is the speed of light, m_e = $9.11 \cdot 10^{-31}$ kg the electron mass, and e = $1.602 \cdot 10^{-19}$ As the elementary charge, h = $6.626 \cdot 10^{-34}$ kg m²/s is Planck's quantum of action and ε_0 = $8.8542 \cdot 10^{-12}$ As/Vm the electric field constant.

Although we have skipped many important details here, at the end of the day Bohr's model of the atom made it possible for the unexplained number R found by Balmer to be expressed by other known constants of nature. By doing so, the number of fundamental constants was reduced by one, and this is *the* epistemological progress provided by atomic and quantum theory. Having explained just one of several constants of nature may not seem a great deal at first sight. Rydberg's constant R, however, was distilled from so many single measurements of atomic spectra that one must consider Bohr's achievement based on Balmer as similar to what Kepler and Newton had accomplished – altogether a dramatic simplification.

WAVES AS PARTICLES, PARTICLES AS WAVES

Someone may have missed the discussion of the wave nature of electrons, which of course was a decisive step on the way to understanding atoms. In his doctoral thesis in 1923, the French physicist Louis Victor de Broglie argued that electrons, like any elementary particle, could also display a wave nature. If, according to Einstein's quantum interpretation of light, a wave sometimes behaved like a particle, then according to de Broglie's creative reasoning, an electron could also behave like a wave. De

2 Large and Small Scale Simplicity: Gravity and the Quantum

Broglie developed a model (how he arrived at this would require closer consideration[6]) in which the wavelength of the electron was given by

$$\lambda = \frac{h}{mv},$$

where m is the mass of the electron and v its velocity. This proposition was convincingly borne out by diffraction experiments of electron beams on crystals by Davisson and Germer in 1927. How does this, certainly revolutionary, insight fit into our scheme of discoveries? Here, the simplification took place before any unexplained numbers could even emerge. If diffraction experiments with electrons had been carried out earlier, poor understanding would have led to a phenomenological description of the outcome. That would certainly have done in terms of free parameters, which would have been regarded as properties of the electron. Instead, with his visionary idea, de Broglie had already explained these potentially unexplained numbers in advance. During that extraordinarily successful phase of physics at the beginning of the 20th century, things often fitted together immediately.

The wave nature of the electrons finally resolved a dilemma inherent in Bohr's atomic model. Being electrically charged particles, orbiting electrons would inevitably radiate energy, the energy loss in turn leading them to crash into the atomic nucleus soon – a fatal inconsistency of the 'little solar systems'. If, however, electrons on a given orbit are regarded as stationary waves, this catastrophe is avoided. At the same time, the quantization of the angular momentum is explained by the fact that only an integer number of waves 'fit' into the electron orbit. If we look at the development of atomic physics as a whole, many interrelated breakthroughs were needed to finally arrive at a convincing picture.

Part I: A Brief History of Physics

INTERWOVEN HISTORY OF SCIENCE

There is one more aspect that must be mentioned. It was a primarily mathematical achievement that led to a consistent form of the atomic model and made wavelike electrons possible in the first place. This important justification goes back to Werner Heisenberg and Erwin Schrödinger, who achieved it by using quite different approaches in 1925 and 1926.

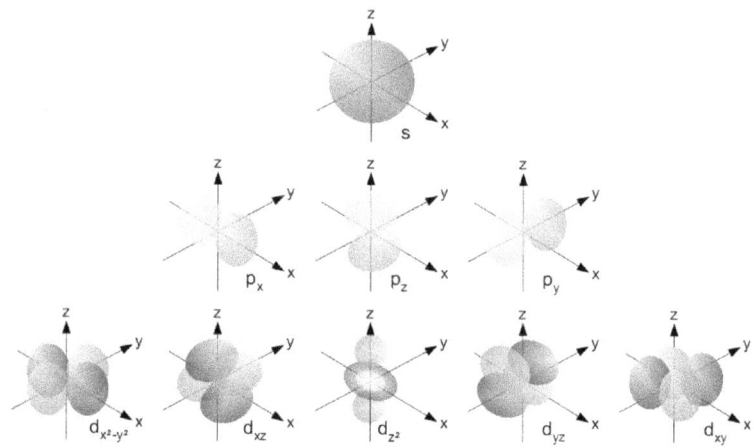

Atomic orbitals whose shape is determined by solutions of the Schrödinger equation. The number of orbitals explains the structure of the periodic system of the elements. For instance, the second shell contains the four orbitals of the first and second row, which can host eight electrons. Correspondingly, there are eight different groups. The third shell (obtained by adding the bottom row) explains the additional elements of higher atomic numbers, in this case metals.

The mathematically consistent formulation of a scientific revolution is often the most difficult and intellectually challenging part, as in the case of Newton's derivation of the elliptic orbits proposed by Kepler. Without such a solid foundation however, a new scientific theory would never prevail.

Nevertheless, the role of the visionary idea must not be underestimated. It requires more creative than technical skills and, in

2 Large and Small Scale Simplicity: Gravity and the Quantum

the case of quantum theory, was achieved primarily by Einstein and Bohr. Bohr was by no means an outstanding mathematician, but he put the pieces of the puzzle together with unique intuition.

If one tries to summarize what fundamental progress consists of, the element of simplification remains the most important one, even if it sometimes appears as a by-product of brilliant math. Condensing a multitude of unexplained numbers into one single constant R, found by Balmer, and the subsequent calculation of R by Bohr that anticipated the explanation of so many later experiments, are the central achievements of quantum mechanics.

With gravitational and quantum theory, we have already touched on the current limits of physical knowledge – on both the large and the small scale. However, before we return to these boundaries, it is important to see how the entire building of modern physics was constructed by successive simplification.

Part I: A Brief History of Physics

3 Heat, Radiation, and Matter: Modern Physics is Emerging

At the end of the 19th century, intense research was carried out on so-called blackbody radiation, which had important technical applications for the newly invented light bulb, inter alia. A particular point of interest was how much radiation a body of a given temperature emitted within a range of wavelengths. Two empirical models were developed from the experimental data: the Rayleigh-Jeans law, which described large wavelengths only, and Wien's displacement law, which also had a limited range of application. Each law contained only one free parameter, which was obtained by fitting the data to the model.

Beyond this rather technical story, however, there is an astonishing visionary idea: since ultimately all materials contain electrically charged particles that move when exposed to heat, it turned out that the radiation characteristics of all bodies could indeed be described by a single formula![1] This discovery in 1900 was due to Max Planck and his mathematical capabilities. It allowed him to identify the above empirical laws as limiting cases of a more general formula, today known as Planck's law of thermal radiation:

$$I(\lambda)d\lambda = \frac{8\pi h f^3}{c^3} \frac{1}{e^{\frac{hf}{kT}}-1}.$$

The law is best illustrated in a diagram that displays the amount of radiation emitted by a blackbody at a given wavelength.

---

[1] Planck's law of radiation is in fact one of the most important findings of modern physics, even if it is often incorrectly applied to gases that cannot emit blackbody radiation. The law is another example of a scientific breakthrough leading to a reduction in the number of constants of nature.

Part I: A Brief History of Physics

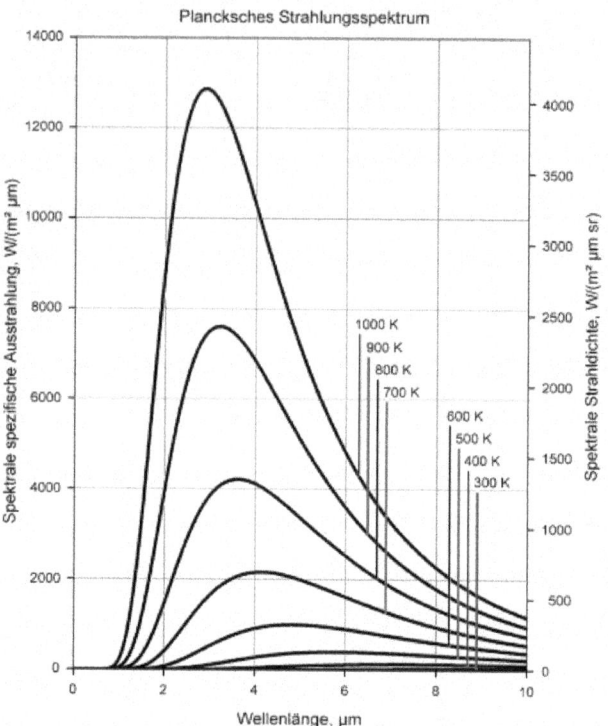

Left: Wavelength-dependent emission of a blackbody according to Planck's law of radiation. The solar spectrum corresponds approximately to a blackbody of 5800 Kelvin.

The emission increases strongly with increasing temperature, and the maximum shifts simultaneously towards smaller wavelengths. It was only much later that Planck, who had initially guessed the correct formula, provided a theoretical justification. He remained overly careful in interpreting the constant h he had introduced, and called it a purely mathematical "auxiliary quantity" without any physical meaning.[1] According to our criterion

---

[1] The German term for auxiliary ('Hilfs-') was the eponym for h.

3 Heat, Radiation, and Matter: Modern Physics is Emerging

of simplicity however, Planck's radiation law reduced the number of constants of nature by one. The two parameters of the laws of Wien and Rayleigh-Jeans became obsolete when h came into existence.

The great importance of h as a constant of nature, however, was recognized by Einstein. At the beginning of the 20th century, the strange results of the photoelectric effect (emission of electrons when light hits a metal) had puzzled many physicists. It was this effect that led Einstein to the formula $E=hf$, which described the energy of 'light quanta', a bold assumption that got the quantum revolution going. As a mathematical formula, the equation $E=hf$ is almost trivial, and it is not immediately obvious which constants of nature it eliminates. However, any other interpretation of the photoelectric effect would have produced poorly understood parameters, while Einstein's formula elegantly solved everything. Thus here, too, the quantum of action h led to a reduction in unexplained numbers in nature, apart from the fact that the concept of light quanta had revolutionary consequences. Although named after Planck, it was Einstein who filled the 'auxiliary constant' with life, providing an interpretation that Planck, of all people, was reluctant to accept. As late as 1913, he publicly mocked Einstein saying that he had "overshot the mark with his speculations".

GOOD AND BAD CONSTANTS OF NATURE?

Perhaps it is time now to reflect upon the 'quality' of constants of nature. The quantum h is certainly considered such a fundamental constant, which sounds positive and connotes importance, while the term 'free parameter' is usually derogatory. This is justified in the sense that 'parameters' are often merely fitted to the data without much mental effort, and consequently no further significance is expected from them. On the other hand, quantities such as h that bear a physical meaning are obviously important messages of nature and often have a 'career' across various research fields. Planck's action quantum h, in particular, has subsequently surfaced in an astonishing number of contexts, which highlights its importance. Even greater is the riddle of the origin of h, which I shall address in the third part of the book.

As a by-product, so to speak, Planck's law of radiation led to another simplification. The total radiation per area emitted by a blackbody is proportional to the fourth power of the absolute temperature T. Independently of Plank, this was already known as the Stefan-Boltzmann law: $w=\sigma T^4$. The constant σ ("sigma") can be determined empirically and represented a generic case of a free parameter, i.e. a 'constant of nature'. With Planck's law it became possible to deconstruct σ – i.e. to calculate it using other constants of nature:

$$\sigma = \frac{2\pi^5 k^4}{15 h^3 c^2}.$$

Once again, the number of constants of nature had decreased by one.

THERMODYNAMIC REVOLUTIONS

Perhaps you have already been trained to recognize free parameters and wondered why the so-called Boltzmann constant k pops up in Planck's formula. The constant k involves a particularly instructive example of a scientific revolution, in which the vision-mathematization-simplification pattern can be identified.

Surprisingly, long after the invention of the thermometer, the origin of heat was still unknown. The visionary idea that heat is nothing other than motion on a molecular level had been formulated by the German physician Robert Julius Mayer in 1842 – a pretty smart conjecture in the day. The courage required to develop such thoughts is hardly appreciated nowadays, because things can be formulated so easily in retrospect. It is remarkable

3 Heat, Radiation, and Matter: Modern Physics is Emerging

how Mayer struggled with the math when developing his theory. In the formula

$$\frac{1}{2}mv^2 = \frac{3}{2}kT,$$

which related the mean kinetic energy of a particle to the temperature T, Mayer had initially forgotten the factor ½. Unfortunately, this hindered the recognition of his achievement. His thoughts were subsequently completed by James Prescott Joule, while the theorist Ludwig Boltzmann from Vienna later used the constant k in the famous formula

$$S = k \log W,$$

which related the amount of heat contained in a physical system to its entropy S (disorder).[I]

If we again think of simplification in physics by reducing the number of constants, then Mayer's formula does precisely that. Prior to its discovery, temperature measurements were, metaphorically speaking, questions to nature that were answered with certain numerical values. But it remained an unsolved problem why a thermometer lying in the blazing sun[II] will display a maximum of 96° centigrade and not, for example, 200°. It was only Mayer's insight that established a connection between the (arbitrary) temperature scale and microscopic kinetic energy. Consequently, the constant k is now no longer considered fundamental, but a definition of temperature. This is correct, but that was exactly Mayer's and Joule's achievement: one constant less.

[I] W stands for the number of microstates of a system, a calculable number.
[II] Again, this can be calculated by equating the incoming solar radiation with the radiation emitted according to the Stefan-Boltzmann law.

Part I: **A Brief History** of Physics

NUMEROLOGY LEADS TO MODERN TELECOMMUNICATION

We have seen that the fields of atomic physics and thermodynamics are closely interrelated, but there are even connections if we consider radiation and the nature of light. Since the Dutch physicist Christiaan Huygens had developed the concept of diffraction in the mid-17th century, it was assumed that light had wave properties. Indeed, it had long been possible to measure the wavelength precisely. However, the nature of light remained unclear until James Clark Maxwell formulated his theory of electrodynamics around 1864. In this theory, separately measurable constants ε_0 and μ_0 appear that quantify the respective strength of the electrical and magnetic interaction. A surprising consequence of Maxwell's equations was that electric and magnetic fields could propagate in empty space, without any electric charges nearby. According to the theory, these waves should propagate at a certain speed.

The German physicists Wilhelm Weber and Rudolf Kohlrausch had already measured this velocity in 1855/56, and it was probably Weber or Kirchhoff who had the visionary idea that light could be an electromagnetic wave.[7] When Heinrich Hertz produced electromagnetic waves in the laboratory for the first time in 1888, he discovered that they actually spread at the speed of light, and the bold hypothesis was spectacularly confirmed. To bring all this into a mathematically consistent form, the entire Maxwellian theory is required,[1] but the revolutionary idea is already contained in the simple formula

$$\varepsilon_0 \mu_0 = \frac{1}{c^2},$$

---

[1] Weber made very significant contributions to this theory, as is clear from Maxwell's treatise that mentions Weber numerous times.

3 Heat, Radiation, and Matter: Modern Physics is Emerging

which reduces the number of constants of nature by one. Instead of three independent constants c, ε_0 and μ_0, only two are left. This is a particularly striking example of how scientific revolutions are characterized by simplification. At the same time, few breakthroughs have had such a lasting impact on civilization as this one.

EINSTEIN'S SIMPLIFICATION

Finally, the constant of nature c plays an essential role in Einstein's theory of special relativity. Einstein realized in 1905 that the measurable speed of light did not change for an observer in motion, which leads to a surprising result: moving clocks run slower, a phenomenon known as time dilation, which is described by the formula $\frac{t'}{t} = \sqrt{1 - \frac{v^2}{c^2}}$. Another consequence is the so-called relativistic mass increase amounting to $\frac{m}{m_0} = \frac{1}{\sqrt{1 - \frac{v^2}{c^2}}}$, a formula that has been convincingly confirmed by experiments in particle accelerators.

The math needed for the formulation of the theory of special relativity is less sophisticated than one might expect. $E=mc^2$, perhaps the most famous formula in physics, is also part of it. Yet where is the simplification in the epistemological sense that I have claimed to be a characteristic of scientific breakthroughs? Which constants of nature or parameters became superfluous as a consequence of Einstein's theory?

AVOIDING THE COMPLICATIONS OF NATURE

This is a special case where the experiments had no time, so to speak, to produce poorly understood parameters, because Einstein's theory was already available as an explanation at the moment they were carried out. Most physical theories are developed

Part I: A Brief History of Physics

after some puzzling results of experiments call for an explanation. The peculiarity of Einstein's achievement was that he developed his theory by pure deduction: only the constancy of c is assumed, all the rest follows from sheer logic. Without Einstein's insights, the advancing technology of particle accelerators in the[I] 1930s would probably have found anomalies such as an increase of mass with velocity. No doubt these results could have been described in a model using some ad hoc parameters.[II] If another theorist had then developed the theory of relativity, the parameters, tentative constants of nature, would have been eliminated by formulas such as $\sqrt{1 - \frac{v^2}{c^2}}$.

Even more concretely, however, the relation $E = mc^2$ has contributed to simplification. The discovery of radioactivity by Henri Becquerel in 1896 showed for the first time that atomic nuclei, and thus chemical elements, are not unchangeable, but could transform into other elements. Ernest Rutherford performed multiple experiments with alpha particles (nuclei of helium) emitted by large nuclei, but it became clear only later that the products of nuclear reactions were lighter than their initial constituents. The immense amount of energy released in such reactions became obvious after the discovery of nuclear fission and its devastating consequences. Nowadays, the amount of energy and the corresponding difference in mass before and after the reaction can, of course, be measured precisely for any reaction.

All these measurements show excellent agreement with the formula $E = mc^2$, which may be called the basis of all nuclear physics. Tested with thousands of individual reactions, it plays a

---

[I] Whether physics would have developed the same way without Einstein is, of course, hard to judge.

[II] When contemplating current high energy physics, which has introduced dozens of arbitrary parameters to describe its experiments, one might suspect that a fundamental theory in the style of Einstein is lacking.

3 Heat, Radiation, and Matter: Modern Physics is Emerging

role in physics similar in importance to Bohr's atomic model describing thousands of spectral lines, or Newton's law of gravity that explains the motion of even more celestial bodies. Einstein's formula $E=mc^2$ thus provides an equally dramatic simplification, recalling our definition based on information theory.

Albert Einstein (1879-1955)

With regard to his most famous formula, Einstein was particularly far ahead of the experiments of his time. Just imagine if the technique of unleashing nuclear reactions had been developed without knowledge of the speed of light! In that case, a 'constant of proportionality' c^2 relating the reaction energy to the mass defect would probably have been 'discovered'. This little historical thought experiment once again shows the extraordinarily successful and dominant role of the constant of nature c in physics. In the last part of the book, special attention will therefore be given to c, as well as to Planck's quantum of action h.

Part I: A Brief History of Physics

DEMOCRITUS IS COMPLETED

In an even broader sense, Einstein's contribution to nuclear physics $E=mc^2$ completed an almost ancient paradigm of physics. After more than two thousand years, the dream of the Greek philosopher Democritus of describing matter in terms of simple building blocks had come true. Democritus' postulate can be seen as the visionary idea behind a long-lasting revolution in science that culminated in modern nuclear physics. Of course, a huge number of scholars in physics and chemistry were involved in this historic endeavor. To begin with, the experimental results involving various substances were overwhelmingly complicated, at least if one wanted to keep track of how many numerical values nature was communicating. Physical properties such as the density of all substances were 'constants of nature' in our sense, as was the energy released in a multitude of reactions among the substances.

The chemists John Dalton and Amedeo Avogadro deserve credit for having quantified atomic weights for the first time. Similar to Johann Jakob Balmer in the atomic spectra, they searched for small integers in the fractional mass ratios and ultimately discovered regularities. While the element of mathematization in these findings may be considered quite primitive, the heuristic obviously led to simplification. A major breakthrough occurred when Italian chemist Stanislao Cannizzaro pointed out the difference between atomic and molecular weight during a conference in Karlsruhe (Germany) in 1860.

Attentive listeners in the audience were Lothar Meyer and Dmitri Mendeleev, who subsequently developed the periodic table of chemical elements. The visionary idea of this partial revolution in 1869 suspected a connection between atomic mass and chemical properties, while the classification into eight groups back then was justified only phenomenologically, by considering

3 Heat, Radiation, and Matter: Modern Physics is Emerging

chemical similarities.[1] The atomic masses of the various elements – undoubtedly constants of nature in our sense – were not yet explained, but an important step in that direction had been taken. Unfortunately, precision was much hampered by nature's whim to create the same chemical elements with different atomic masses – so-called isotopes. This was realized much later, in 1911, by the Belgian chemist Soddy.

A SINGLE MASS

It was soon established that the nuclei of a given chemical element contained the same number of protons, but a different number of uncharged neutrons (which have approximately the same mass). By classifying atomic nuclei in this way, Democritus' vision already shone through: their masses were approximate multiples of the atomic mass unit u, named after the pioneer John Dalton. When the first nuclear reactions were conducted in the 1930s, Einstein's formula, already well-known at that time, was extensively validated and found to be in precise agreement with all the results. All in all, the results of atomism represent a brilliant confirmation of Democritus' old idea and constitute an essential part of mankind's knowledge.

The masses of all known chemical elements were thus found to be multiples of the mass of the proton (electrons in the shell hardly contribute to mass). The enormous simplification brought about by this long-lasting revolution in atomic theory is evident: Regarding mass, there is effectively just one fundamental constant, the mass of the proton m_p with the numerical value $1.6726 \cdot 10^{-27}$ kg.

---

[1] Only the Schrödinger equation found in 1925 and its solutions justified this classification and thus completed the mathematical formulation of the periodic sytem. Of course, this is another a great achievement of quantum theory.

Part I: A Brief History of Physics

The proton mass obviously plays an important role in nature, but the question why the proton is so heavy remains. This will be discussed in more detail in Chapter 6. Another riddle is the mass ratio of proton and electron $m_p/m_e=1836$, about which Paul Dirac had racked his brains all his life. In contrast to the mindset of today's physicists, even the existence of two particles contradicted Dirac's idea of simplicity in nature.

In the above discussion we have so far bypassed the electric charge of atomic nuclei, which played an important role in the history of atomism. In 1923, experiments conducted by the American physicist Andrew Millikan had proved the astounding fact that nature allows the electric charge to occur only in certain quanta (namely the charge of the electron), even if no deeper explanation for this mystery is known to this day. Soddy was awarded the Nobel Prize in Chemistry in 1921 for his discovery that chemical properties were actually determined not by the mass of the atomic nucleus (since there are different isotopes anyway), but by its charge. However, prior to Soddy, another great thinker had already suggested this groundbreaking rule – Niels Bohr!

CHARGES EXPLAIN CHEMISTRY

Let us pause and contemplate for a moment the tremendous progress in understanding that was achieved by the collective efforts of physicists and chemists over the centuries: almost one hundred stable chemical elements, starting with hydrogen, helium, lithium, etc., can readily be explained by 1, 2, 3... protons in the nucleus! Again, the enormous simplification of natural

3 Heat, Radiation, and Matter: Modern Physics is Emerging

phenomena needs no further comment. It is reflected in the existence of the elementary charge e=$1.602 \cdot 10^{-19}$ Coulomb[I], a constant of nature that will also be discussed later.

This review of the history of physics would be significantly incomplete without mention of the unification of electricity and magnetism at the beginning of the 19th century. After occasional reports of magnetic needles being affected by thunderstorms, a link between electricity and magnetism was widely suspected – the visionary idea of unification was literally in the air here, yet presented itself in an unexpected manner.

In 1820, the Danish physicist Hans Christian Ørstedt demonstrated that a magnetic needle oriented itself perpendicular to a current flowing through an electric conductor; however, the same experiment had already been published in 1802 by the Italian physicist Gian Domenico Romagnosi. When the effect became known in Central Europe, it was above all the preeminent British scholar Michael Faraday who performed systematic investigations on the subject, while the Frenchman André-Marie Ampère succeeded in formulating the relevant mathematical laws.

However, dealing with the units commonly used at the time would be quite cumbersome, and the first results of this revolution were not yet expressed quantitatively. Nevertheless, it is obvious that the unification of formerly separated electric and magnetic phenomena was a simplification that reduced the number of arbitrary constants. In modern notation, this is reflected by the convention that the magnetic field constant μ_0 is no longer a 'real' constant of nature, but has been incorporated into the definition of electric current. Consequently, the exact[II] value $4\pi \cdot 10^{-7}$ Vs/Am has been assigned to μ_0.

[I] The unit of charge.
[II] In the 2019 update of SI units, this century-old convention was overturned again, yet this has no relevance for our discussion here.

Part I: **A Brief History** of Physics

AT THE SUMMIT OF SIMPLICITY

Around 1930, physics was at the summit of its glory. Being able to describe all the results known until then with only a few constants of nature represented a methodological frugality that reflected the great achievements of the natural sciences. At this point it is useful to summarize the most important breakthroughs with particular attention to the vision-mathematization-simplification pattern. In the process, the far-reaching unification of physical theories also becomes evident.

Players	Year	Vision	Formula	Obsolete
Kepler	1600 1610	Sun at the center	Kepler's Laws	Epicycle
Newton	1687	Earthly and celestial gravity	$g = \dfrac{GM}{r^2}$	G
Balmer	1885	Mathematics in atomic spectra	$1/\lambda = R\left(\dfrac{1}{2^2} - \dfrac{1}{3^2}\right)$,	$\lambda_1, \lambda_2, \ldots$
Weber, Hatamura, Bohr	1904 1913	Atoms as Solar System h as angular momentum	$R = \dfrac{m_e e^4}{8c\, \varepsilon_0^2 h^3}$	R
Planck	1900	Uniform radiation law	$I(\sigma, \lambda) = \ldots$	Wien, Rayleigh-Jeans
Maxwell, Weber, Ampère	1864	Unification Electricity Magnetism	Maxwell's equations	μ_0
Hertz, Weber	1888	Light is elmag. wave	$1/c^2 = \varepsilon_0 \mu_0$	ε_0
Mayer, Jewel	1842	Heat is kin. energy	$\tfrac{1}{2} mv^2 = kT$	k
Democritus Dalton Mendeleev Einstein	500 BC - 1930	Matter from elementary building blocks	Schrödinger equation, periodic table etc.	Atomic masses
Einstein	1905	Constant c	$E=mc^2$, $t'/t = \ldots$	…

3 Heat, Radiation, and Matter: Modern Physics is Emerging

Let us remember that constants of nature play an essential role in physics. The appearance of a new constant of nature, say the elementary charge e, can be a significant discovery. Sometimes, researchers have stumbled across new quantifiable properties that nature liked to communicate. However, as we have seen in these two chapters, the vast majority of these numbers can be explained by the efforts of physicists over the centuries.

Each explanation in which a constant of nature became obsolete signified a major accomplishment – another secret that had been wrested from nature. On the other hand, all major breakthroughs were accompanied by a reduction in the number of fundamental constants. The remaining constants of nature must therefore be regarded as particularly important; 'hardened' numbers that have so far resisted any attempts at explanation. These numbers therefore require special attention, because it is within them that the secrets of nature are hiding. This is where the future revolutions of physics will take place.

HOW MANY CONSTANTS OF NATURE ARE THERE?

It is useful here to give a short overview of fundamental constants. We have the size of the proton $r_u = 0.84 \cdot 10^{-15}$ m and its mass $m_p = 1.6726 \cdot 10^{-27}$ kg. Instead of considering the mass of the electron m_e, which could also be regarded fundamental,[1] we chose the mass ratio $m_p/m_e = 1836.15 ...$, obviously an unexplained number. Likewise, we have already discussed the mysterious numerical value

$$\frac{e^2}{2hc\varepsilon_0} \approx \frac{1}{137},$$

---

[1] In contrast to the size of the proton, there is no measurement of the size of the electron r_e independently from theory, which is why I do not list it as a natural constant.

Part I: A Brief History of Physics

the so-called fine structure constant, another pure number of unknown origin. Richard Feynman commented on it as follows:

> *It's one of the greatest damn mysteries of physics: a magic number that comes to us with no understanding by man. You might say the 'hand of God' wrote that number, and 'we don't know how He pushed his pencil.' We know what kind of a dance to do experimentally to measure this number very accurately, but we don't know what kind of dance to do on the computer to make this number come out, without putting it in secretly (...). All good theoretical physicists put this number up on their wall and worry about it.*[8]

For the time being, I would like to exclude the mass of the neutron m_n from our collection of fundamental quantities. The neutron weighs only slightly more than the proton and is an unstable particle. For at the beginning of any discussion the question would arise why nature needs radioactive particles at all (the neutron decays on average after about 15 minutes). To what extent would a physics without radioactivity contradict fundamental laws of nature? This puzzle is unsolved, so I prefer not to bother with the confusing variety of decaying particles; although the half-life of the neutron t_n is actually a number that ought to be explained. Incidentally, half-life anomalies are a very interesting topic, although little attention is paid to them in the current paradigm.[9] More on this later.

The number of 'electrical' constants has already been reduced by the abovementioned revolutions triggered by Faraday, Maxwell and Hertz, and all remaining constants are therefore contained in the definition of fine structure constant $\frac{1}{137} \approx \frac{e^2}{2hc\varepsilon_0}$. Could that number ever be calculated, the issue of the electrical constants would be solved, presumably accompanied by deep insights. To complete the count of fundamental constants so far, the

3 Heat, Radiation, and Matter: Modern Physics is Emerging

speed of light c, Planck's quantum of action h, and the gravitational constant G must be added as free parameters.

Anyone who has followed modern physics since 1930 will have noticed the large number of newly discovered particles. They have been used to measure a great many quantitative properties of nature which, according to our definition, should be called constants of nature: think of muons, pions, various flavors of quarks and neutrinos, or even W-, Z- and Higgs-bosons. However, since fundamental progress has always been accompanied by a reduction in the number of such parameters, it is reasonable to conclude that the standard model of particle physics is a dead end, much like the medieval epicycles.[10] Since we are interested in elementary insights, I will therefore not include those models from after 1930 in our discussion, since back then physics got along with the fewest constants of nature. For without having solved the puzzles that existed at the time, efforts to understand a multitude of new parameters would be doomed to fail anyway.

Part I: A Brief History of Physics

4 Cosmology Explains the Gravitational Constant

In the midst of a successful era of physics, when the understanding of the microworld had reached new heights, a new window on the physical universe opened up: cosmology. After much debate on the subject, the American astronomer Edwin Hubble confirmed in 1923 that the blurred structure in the sky in the Andromeda constellation was in fact another galaxy similar to our own Milky Way.

It is no wonder that astrophysics and cosmology have flourished since then, though I cannot set out the story in full detail. In essence, however, observational cosmology led to measurements that must be regarded as free parameters or constants of nature. Despite these fascinating new data – I may anticipate – the understanding of these cosmological parameters is no further advanced than in the microworld of atoms.

Incidentally, this has nothing to do with the fact that Einstein extended his theory of special relativity of 1905 in the following years to a general theory of relativity that included gravity.[1] Simply because the constants of nature which observational cosmology should deliver one day were not known at the time! It wasn't until 1929 that Hubble discovered that the light of distant galaxies was shifted towards the red end of the spectrum. Since then, this has been interpreted as an expansion of the universe; in any case, Hubble's observations are evidence for a finite age and finite size of the universe.

---

[1] Of course, this theory was confirmed by measuring values, which again had no time to become misunderstood parameters.

Part I: A Brief History of Physics

Before we look more closely at the cosmic evolution, let me once again mention Paul Dirac, who in 1937 was the first to point out a possible connection between the physics of the cosmos and the physics of atoms.[11] From our epistemological point of view, this is a visionary idea, the intriguing consequences of which I will discuss in depth later.

Unfortunately, in the past decades cosmological research has developed in a similar fashion to particle physics, despite fantastic observations. Surprising data, or even contradictory data, have usually been 'explained' by new free parameters, leading to increasingly complicated models. The analysis of these models, in particular the concepts of 'dark matter' and 'dark energy', is rather pointless, if not detracting when one is searching for fundamental laws. Therefore, I shall focus instead on the important constants of nature provided by cosmology.

In our technical sense, these are quantitative measurements, such as the age of the universe t_u, for which estimates have existed since the 1930s. If one assumes that light signals have been propagating since its beginning (commonly called the 'big bang'), the size of the cosmos[1] can be estimated by $R_u = c \cdot t_u$, the product of the age of the universe and the speed of light. Since this equation makes one constant superfluous, for the sake of simplicity I will only consider the radius of the universe R_u (approximately 10^{26}m) in the following. Determining the mass of galaxies and the average density of galaxies in the universe[12] is possible, although reasonably accurate measurements were available only much later than 1930. Rather than determining the precise value (which is a matter of debate anyway) we are interested in

---

[1] Instead of the age t_u, the so-called Hubble constant $H_0 = 1/t_u$ is often considered. The radius of the horizon R_u is also a model-dependent value. However, we are interested in orders of magnitude, on which the distinction has no effect.

4 Cosmology Explains the Gravitational Constant

the theoretical possibility of estimating the mass M_u of the universe (about 10^{52} kg) within the current horizon R_u (about 10^{26} m). Dividing this mass by the corresponding volume of a sphere yields an approximation of the average density of the universe, roughly corresponding to one atom per cubic meter.

WHERE DO WE GO FROM HERE?

Hence, if we summarize the state of particle physics around 1930 in a somewhat idealized way,[1] and in addition, consider the messages from the cosmos, the simplest description of nature still requires at least nine different constants, namely G, h, c, M_u, R_u, m_p, r_p as well as the two pure numbers 137.036 and 1836.15...

It has become clear from the previous chapters that fundamental progress can only be achieved by further reducing the number of constants of nature. However, taking into account the revolutionary breakthroughs in physics that have happened in the past (none is visible in the past 100 years), expectations look gloomy. This is all the more true since physics has apparently failed in recent decades to crack one of those nine obviously 'tough nuts'.

However, the majority of physicists unfortunately overlook the fact that the work of some visionary thinkers has already reduced the number of those constants, and further explanations seem within reach. Apart from Paul Dirac mentioned above, these considerations go back in large part to Albert Einstein. Perhaps his most important line of reasoning about general relativity in 1911 has remained practically unknown until today. I have given a detailed account of this idea and its consequences in my book *Einstein's Lost Key*, in which the reader can scrutinize both the experimental evidence and the historical development. Here,

---

[1] Measuring values such as the muon mass m_m=206.77... m_e are certainly intriguing, but they offer little help from the perspective of natural philosophy.

however, I shall limit the discussion to the possible elimination of the gravitational constant G and its revolutionary consequences.

THE THINKER FROM VIENNA

The visionary idea of explaining the gravitational constant goes back to the Viennese philosopher and physicist Ernst Mach. In his famous 1883 treatise *Mechanik,* Mach had expressed fundamental, though almost heretical ideas on gravity, while criticizing Newton. Newton had justified his postulate of absolute space (this is what is questioned here, too) with an experiment of a bucket in which the water rises up on the walls when the bucket rotates. Newton claimed that this proved the existence of an absolute, motionless space in which there is no acceleration. Mach, however, came up with the following profound objection: *"Nobody can say how the experiment would turn out if the bucket walls became increasingly thicker and more massive, eventually several miles thick..."* Obviously, he was suggesting that the distant masses in the universe could be the cause of the inertia.

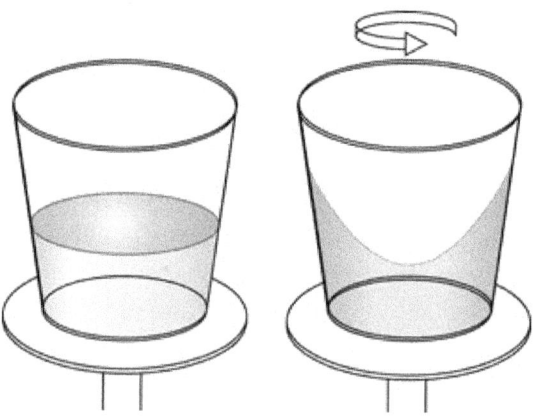

Newton's bucket experiment: Standing bucket (left) and rotating bucket (right) in which the water level rises up on the walls.

4 Cosmology Explains the Gravitational Constant

In a further visionary leap of imagination, Mach contended that both the inertia and gravity of a mass are of the same nature – an idea that Albert Einstein 25 years later would call equivalence principle, the basis of his theory of general relativity.[13]

Ernst Mach (1838-1916)

As a consequence, Mach argued that the origin of gravity must lie in the presence of all other masses in the universe. Since then, this profound thought has become known as *Mach's principle*, but its importance is grossly underestimated. In hindsight, it could be regarded as tragic that Ernst Mach – 40 years before observational cosmology was born – did not have the opportunity to see his conjecture confirmed. For it did not become clear until the end of the 1930s that the mass and size of the universe are fascinatingly related to the gravitational constant and the speed of light c. This coincidence should actually have helped Ernst Mach's ideas to gain currency. In fact, it is sensational that the following formula holds in good approximation:

$$G \approx \frac{c^2 R_u}{M_u},$$

63

where c is the speed of light, M_u and R_u the mass and radius of the universe, and G the gravitational constant.

In current mainstream cosmology, this coincidence is called 'flatness' and is associated with the concept of 'cosmic inflation', which, like many modern ideas, has no explanatory value whatsoever. Paul Dirac, on the other hand, had already wondered about this coincidence in 1938. Moreover, Erwin Schrödinger had speculated about the above relation in a far-sighted article[14] on cosmology as early as 1925, even before Hubble's measurements. Schrödinger was led to the idea because he had noticed that the expression for the gravitational potential, GM/r, had the units of a squared velocity, i.e. m^2/s^2. This led him to the bold assumption that the total gravitational potential of the universe was equal to c^2. As has happened frequently before, considering physical units led Schrödinger to deep insights. Therefore, it makes sense to let units serve as a guiding principle for exploring the possible progress of physics.

8. Die Erfüllbarkeit der Relativitätsforderung in der klassischen Mechanik; von E. Schrödinger.

Gegen die klassische Punktmechanik mit Zentralkräften, deren Grundlagen in klarster Form von L. Boltzmann[1]) herausgearbeitet wurden, ist bekanntlich schon von E. Mach[2]) der Einwand erhoben worden, daß sie der vom erkenntnistheoretischen Standpunkt sich aufdrängenden Relativitätsforderung nicht genüge: ihre Gesetze gelten nicht für *beliebig*

Schrödinger's original publication from 1925

EINSTEIN'S GREATEST IDEA

Still earlier, in 1911, Albert Einstein, adopting a completely different line of reasoning, had incorporated this very same idea into an equation that was the beginning of the development of

4 Cosmology Explains the Gravitational Constant

general relativity. The basic rationale of general relativity consists in the equivalence principle anticipated by Mach: it states that inertial mass, defined by resistance to acceleration, and weight, which causes attraction, are basically the same. Though we are quite used to this fact from everyday experience, it remains utterly surprising from a fundamental perspective.

Einstein formulated the principle in a slightly different way: an accelerated reference system cannot be distinguished from gravitational field. In fact, it is hard to determine, by measurement, whether you are in a rocket-boosted spacecraft or in a dark cellar on Earth. In both cases, one feels a force pulling downwards.

It follows directly from this comparison that light rays are deflected by gravitating bodies.[15] Einstein explained this straightforwardly by assuming that masses caused the speed of light to decrease in their vicinity. As in conventional optics, this leads to a curvature of light rays. To quantify his idea of variable speed of light, Einstein developed a formula in which the coincidence regarding G was already contained, but not in an obvious way.

daher die Lichtgeschwindigkeit c in einem Orte vom Gravitationspotential Φ durch die Beziehung

$$(3) \quad c = c_0 \left(1 + \frac{\Phi}{c^2}\right)$$

gegeben sein. Das Prinzip von der Konstanz der Licht-

Excerpt from Einstein's article 1911

But why should the speed of light be variable in first place? First, I should clear up a possible conceptual misunderstanding. Einstein grounded his 1905 theory of special relativity on the 'constancy' of the speed of light, an argument that is still valid. For the measured speed of light does not change no matter how the *observer* moves, i.e. it is constant with regard to a change of the reference system. But since we are talking about just one point at just one instant, that kind of 'constancy' does not at all contradict the concept of a speed of light that depends on space and time. It is more of a linguistic paradox than a real problem.

It can only be regarded as an unfortunate whim of history that Einstein at that time could not have had the slightest idea about the true size of the universe (galaxies were not yet discovered!). Had he been aware of this, the relation $\frac{c^2}{G} \approx \frac{M_u}{R_u}$ would inevitably have attracted his attention. However, without being able to integrate his formula into the broader context of Mach's principle, and due to another miscalculation, he abandoned his idea in the following years in favor of a different formulation of the theory of general relativity that is widely accepted today. I cannot rehash here all the ensuing pitfalls and confusions of cosmology that continue to this day, but from our methodological perspective it is clear that the numerical coincidence $\frac{c^2}{G} \approx \frac{M_u}{R_u}$ alone is sufficient motivation to pursue Einstein's original idea.

INDEPENDENT REDISCOVERY

The surprising equality has inspired several thinkers to form theories. The British-Egyptian cosmologist Dennis Sciama published[16] a fundamental essay on the subject in 1953, but above all, it was American astrophysicist Robert Dicke who rediscovered[17] and crucially improved the variable speed of light formulation of general relativity in 1957, without even knowing about Einstein's approach. The simple relation $G = \frac{c^2 R_u}{M_u}$ suffers from the fact that it does not reveal a physical mechanism. However, according to Dicke's calculation, the well-known quantity of gravitational potential (as previously assumed by Schrödinger and Sciama) is linked to c^2, a beautiful realization of Mach's principle. Consequently, one may write:[18]

$$G \sum \frac{m_i}{r_i} = \frac{1}{4} c^2,$$

4 Cosmology Explains the Gravitational Constant

where Σ means taking the sum over all masses m_i in the universe, divided by their respective distance r_i. Dicke corrected the error that had led Einstein to abandon variable speed of light in 1912. Moreover, he was the first to show that the variable speed of light, like the conventional form of general relativity, describes all known tests. For that reason, it was necessary to accompany c^2 by a factor ¼.

At this point, the reader may wonder why this discovery did not make a big splash in 1957. Although Dicke emphasized that this result was 'highly satisfactory' in view of Mach's principle, he failed to point out that, as a consequence, G could be eliminated – perhaps because he was unaware of the epistemological relevance of such a simplification. When studying the history of science, however, it also becomes clear that it is not always reason, logic, and objective facts that determine which theory is accepted by the majority of practitioners. Rather, there is a lot of sociology and psychology at play that determine the 'established' models.

Einstein's original idea of describing the deflection of light rays by a variable speed of light (instead of by a curved space), improved by Dicke's theory, may rightly be called simpler, if we talk about of accessibility. However, there has been a long-standing mystification of Einstein, in the aftermath of which it is considered almost suspicious if a physical theory can be expressed in an understandable way. The fact that the variable speed of light version of general relativity is equivalent to the conventional geometric form as far as observations are concerned has been demonstrated by a whole series of researchers[19], but sinks only slowly into the consciousness of physicists.

Part I: A Brief History of Physics

IN THE BEGINNING THERE WAS COINCIDENCE

Incidentally, in general relativity, Einstein used the quantity $\kappa = \frac{8\pi G}{c^4}$ instead of Newton's gravitational constant G. This actually turns out to be more practical, and thus in the following instead of

$$\frac{c^2}{4G} = \sum \frac{m_i}{r_i} \text{ we may use the identity } \frac{1}{c^2} = \frac{\kappa}{2\pi} \sum \frac{m_i}{r_i},$$

which is even more fundamental when thinking about mathematical generalizations.[20] However, with respect to observational data,[I] this does not differ from Dicke's theory.

Starting from $\frac{1}{c^2} = \frac{\kappa}{2\pi} \sum \frac{m_i}{r_i}$ with $\kappa = \frac{8\pi G}{c^4}$, Newton's law of gravitation is easily deduced. Since, according to Dicke's formula, the gravitational potential φ is ¼ c^2, taking the gradient yields the local acceleration

$$g = -\nabla\varphi = -\nabla \tfrac{1}{4} c^2 = -\nabla \frac{2\pi}{\kappa \sum \frac{m_i}{r_i}}.$$

The Operator ∇ "Nabla" means gradient, a spatial derivative. By applying the chain rule of differentiation one obtains

$$g = \frac{2\pi}{\kappa \left(\sum \frac{m_i}{r_i}\right)^2} \sum \frac{m_i}{r_i^2} = \frac{c^2}{4 \sum \frac{m_i}{r_i}} \sum \frac{m_i}{r_i^2} = G \sum \frac{m_i}{r_i^2}.$$

Evidently, Newton's inverse-square law appears. In a second step, the term $\frac{2\pi}{\kappa \sum \frac{m_i}{r_i}}$, being equivalent to $\frac{c^2}{4 \sum \frac{m_i}{r_i}}$, is identified with the gravitational constant G, as Sciama and Dicke had suspected.

[I] It is interesting to note that there are other possibilities. One could even use an arbitrary function of the aforementioned sum, although there is no good theoretical argument for doing so.

4 Cosmology Explains the Gravitational Constant

If one carries out the sum by assuming a constant density of the universe,[1] the relation observed earlier is obtained once again: $\kappa = \frac{4\pi R_u}{3 M_u c^2}$ or rather $G = \frac{c^2 R_u}{6 M_u}$, which concretely eliminates the gravitational constant.

This calculation also determines the mass M_u of the universe, which may seem daring, since there are, of course, no accurate astrophysical measurements. However, we are in a situation similar to that of the mass of the Earth, for which only rough estimates can be obtained by geophysical methods. Moreover, the Earth's mass is determined by G.[21] So we rely on theory whenever we peer 'inwards' and 'outwards' from our planet.

4. Über den Einfluß
der Schwerkraft auf die Ausbreitung des Lichtes;
von A. Einstein.

Die Frage, ob die Ausbreitung des Lichtes durch die

Einstein's 1911 article in the journal Annalen der Physik

However, I should point out here that the model involving variable speed of light does not just consist in a coincidence about G that one might call speculative. There is a consistent mathematical theory that quantitatively agrees with the numerous tests of general relativity. As a consequence, this discovery should remind everybody to take such numerical coincidences seriously, even before the corresponding theory has been worked out. For without inspiration by such hypotheses, the majority of breakthroughs in physics would not have occurred: think, for example,

---

[1] Of course, this assumption holds true only approximately. In fact, there could be variations of the gravitational constant due to the spatial distribution of matter. On the other hand, a *temporal* variation of G, which Dirac had predicted in 1938 (perhaps somewhat hastily), is not necessarily observable. For more details, see *Einstein's Lost Key*.

of the coincidences considered by Wilhelm Weber or Johann Jakob Balmer.

THE BIG STEP INTO THE UNIVERSE

Unfortunately, decades before the internet age, Mach, Einstein, Schrödinger, Dirac, Sciama and Dicke could not recognize the common denominator of their ideas, as is possible today. From an epistemological point of view, what counts most is that the gravitational constant G had become obsolete. One needs to put this into historical context, even if this insight is not yet generally accepted. Isaac Newton had the bold idea of linking terrestrial gravity, quantified by local gravity g, to celestial bodies, and this led to the explanatory equation $g = \frac{GM}{r^2}$. More precisely, however, local gravity g is determined by all masses, so that one would have to write (vectors $\vec{e_i}$ because forces point to a direction)

$$g = G \sum \frac{m_i \vec{e_i}}{r_i^2}.$$

Of course, distant masses with large r_i are practically imperceptible. Indeed, g and many other arbitrary numbers were explained in this way and traced back to the constant G, whose origin however remained unknown. Just as Newton unified terrestrial and celestial gravity, Mach dared to leap from the solar system to the universe. The formula $\frac{c^2}{4G} = \sum \frac{m_i}{r_i}$ explains the constant G in an analogous way, just as Newton's law of gravity had explained g. Interestingly enough, the distances r_i from gravitating masses in the denominator appear only linearly, instead of the squares in Newton's law, a peculiarity that made Mach's principle much harder to detect.

Part II: The End of Space and Time

"Don't get involved in partial problems, but always take flight to where there is a free view over the whole single great problem, even if this view is still not a clear one."

Ludwig Wittgenstein

Part II: The End of Space and Time

5 The Cosmos Without Expansion: A World of Variable Scales

After the explanation of the gravitational constant G, in this chapter I will describe a further step of simplification. The visionary idea in general goes back to the already mentioned Paul Dirac, who in 1937 had discovered an intriguing coincidence between the size of the proton and the size of universe. In 1938, he elaborated on this idea in an article in the *Proceedings of the Royal Society*, presuming[22] *"some deep connection in Nature between cosmology and atomic theory"*. Like Einstein, Dirac had apparently spent many years thinking about the unification of gravitation and electromagnetism. But it was Dirac who did so from the perspective of fundamental constants. In the hydrogen atom, nature's simplest stable structure, the constituent proton and electron are held together by electric force. Yet their gravitational attraction, however small it may be, can also be calculated theoretically. Dirac noticed that the ratio of the two forces

$$\frac{F_e}{F_g} = \frac{e^2}{4\pi\varepsilon_0 G m_p m_e} \approx 2{,}3 \cdot 10^{39}$$

was an incredibly huge number with almost 40 digits. Dirac had always been fascinated by pure numbers appearing in nature. Since he took it for granted that it was the task of a theoretical physicist to calculate such numbers, he was shocked by its magnitude. Any attempts to produce it with reasonable math seemed hopeless at first sight. Dirac was electrified by the first estimates of the size of the cosmos when it turned out that the ratio of the visible horizon R_u and the radius of the proton[1] r_u resulted in a

---

[1] The current measurement is $0.84 \cdot 10^{-15}$ m, but this is still based on certain model assumptions that are not supported by a thorough understanding.

similarly large number of about 40 digits. The coincidence between these microscopic measures and those of the universe is indeed striking.

YET ANOTHER COINCIDENCE

All the more incomprehensible is the way that Dirac's hypothesis is utterly disregarded by today's 'leading' physicists; it is frequently dismissed as 'playing with numbers'. Most physicists are unaware of the fact that Dirac had discovered a second, equally striking relation connected to the above ratio, which makes a random coincidence extremely unlikely. After the first estimates of the total mass M_u of the universe in the 1930s, Dirac divided M_u by the mass of the proton m_p, thus arriving at a ballpark figure of the number of particles in the universe, 10^{78}, the square of that other mysterious number 10^{39}! This was completely mysterious, but lent substance to the first observation.[I] For decades, Dirac's second observation has defied all attempts at explanation. In particular, it seems to jeopardize all established cosmological models. For normally the number of particles should be proportional to the volume, i.e. the third power of the linear dimensions of the cosmos, not just to the second. All the more remarkable, however, is that Dirac's cosmology ultimately follows from Einstein's pivotal idea of a variable speed of light back in 1911, if one consequently applies Dicke's formulation of 1957.

In view of the goal of reducing the number of unexplained constants of nature, I will limit the presentation to what is necessary for understanding the key concepts.[II] The mathematical

---

[I] As Dirac noted, factors such as the fine structure constant 1/137 may still come into play. In any case, the order-of-magnitude coincidence is remarkable.
[II] More detail and motivation can be found in Chap.10 of *Einstein's Lost Key*.

5 The Cosmos Without Expansion: A World of Variable Scales

model that arises is not really hard to understand, but requires some brief context.

IN THE BEGINNING THERE WAS LIGHT

> *How can we know that one second today is as long as one second yesterday?* – Julian Barbour

The speed of light c is reduced due to the presence of masses at a distance. This proposition is in agreement with all tests of general relativity, but so far it can be understood as a purely spatial variation of c. However, it is obvious that the visible horizon of the universe grows every day due to the simple fact that light spreads. Hence, the mass contained in that horizon increases as well and consequently, the speed of light must decrease with cosmological time, which is however not visible in laboratory environments. This is because, in addition to the speed of light, the corresponding time and length scales are also variable, which would neutralize any direct measurements of the speed of light c.[1] The elementary formula

$$c = \lambda f$$

applies, where λ is the wavelength and f the light frequency. When c decreases, the wavelengths λ and the frequencies f must become smaller as well. To be in agreement with the tests of general relativity, the relative changes of λ and f must be equal. If, for example, c drops to 96 percent of its value, both λ and f decrease to about 98 percent.

This is true for all light waves absorbed and emitted by atoms, and we must therefore imagine living in a world of variable scales, whose slow change, however, remains hidden from direct

---

[1] The variability of the speed of light, of course, manifests in an indirect manner through the deflection of light by masses.

measurement. When Robert Dicke considered this model in 1957, he associated the change in wavelength λ with the cosmic redshift of light emitted by distant galaxies. When analyzing Maxwell's equations (which govern light waves), he found that for light propagating in the universe, the decrease of c in the above formula is caused *solely* by a decrease in the frequency f, while the wavelength λ remains the same. However, this has a dramatic consequence: if such propagating light reaches atoms at cosmic distances whose reference wavelengths have decreased during the travel time, its wavelength λ appears relatively longer at the moment of arrival. This is the cosmic redshift first observed by Edwin Hubble!

> **COSMOLOGY**
>
> The cosmological principle was taken to be a fundamental assumption of the theory. Namely, from any fixed position point of a Newtonian frame the universe is assumed to be on the average uniform. This implies that matter is on the average fixed in position relative to the Newtonian coordinate frame, for motion would

Dicke's somewhat awkwardly phrased statement about a non-expanding universe in his 1957 paper.

THE UNIVERSE IS NOT EXPANDING

The mind-blowing consequence of Dicke's idea is that the famous redshift of light from distant galaxies should not be interpreted as motion and thus not as an expansion of the universe (for which there has never been a satisfactory explanation anyway). In Dicke's model, the cosmic redshift is rather a consequence of a basic property of light, namely that it spreads. Hence, there is no material expansion of the universe, but only a growth of the region we can see, i.e. the horizon. The speed of light c is nothing other than the rate of growth of that horizon. In turn, this cosmological measurement is linked to the microscopic properties of

5 The Cosmos Without Expansion: A World of Variable Scales

light, to which $c = \lambda f$ applies. This link from the cosmos to light can be seen as an extension of Dirac's visionary idea that already pointed to the right framework of mathematical formulation, since Dicke had already calculated the variability of all measuring scales (which are summarized in the table below).

However, there are two further satisfying aspects to examine: there is no need to wonder about the origin of the mysterious 'expansion' of the cosmos, which turns out to be an illusion, nor does one have to invoke absurdities such as *creatio ex nihilo,* as is required in the conventional Big Bang model. Once again, it is hard to understand why this alternative explanation of the redshift – actually the first explanation ever – has remained so little known until today. Unfortunately, Dicke had chosen a dull title for his publication and mentioned his revolutionary result of the non-expansion only casually on one of the last pages.[23]

But how can Dicke's model reduce the number of constants of nature? To understand the following, we need to delve into some mathematical notation, which will allow things to be properly illustrated. The basis of Dicke's considerations is that the visible horizon expands at the speed of light. Since in this model all time and length scales are variable, it is appropriate to define formally absolute scales in which those changes can be quantified.

THE BIG FLASH THEORY

As Dicke had realized, mathematical consistency required that the speed of light decreases with the root of absolute time, in formal notation:

$$c \sim t^{-\frac{1}{2}} = \frac{1}{\sqrt{t}}.$$

Part II: The End of Space and Time

This has a number of consequences that are summarized[1] in the following table.

Size		units		Epoch	Example	
Absolute time	t	s	t	10^{52}	10.000	↑↑↑↑
Speed of light	c	m/s	$t^{-1/2}$	10^{-26}	1/100	↓↓
Wavelength	λ	m	$t^{-1/4}$	10^{-13}	1/10	↓
Frequency	f	1/s	$t^{-1/4}$	10^{-13}	1/10	↓
Time step	Θ	s	$t^{1/4}$	10^{13}	10	↑↑
Velocity	v	m/s	$t^{-1/2}$	10^{-26}	1/100	↓↓
Acceleration	a	m/s²	$t^{-3/4}$	10^{-39}	1/1000	↓↓↓
Inertial mass	m	kg	$t^{3/4}$	10^{39}	1000	↑↑↑
Horizon Universe	R	m	$t^{1/2}$	10^{26}	100	↑↑
Measured horizon	R'	s	$t^{3/4}$	10^{39}	1000	↑↑↑
Measured time	t'	s	$t^{3/4}$	10^{39}	1000	↑↑↑
Volume Universe	V	m³	$t^{3/2}$	10^{78}	1.000.000	6↑
Number of particles	N	–	$t^{3/2}= t^2$	10^{78}	1.000.000	6↑
Mass Universe	M	kg	$t^{9/4}= t^3$	10^{117}	10^9	9↑

Variability of physical quantities in Dicke's model. The exponents are rounded to integers, so the exponent over t is closer to 53 than to 52. Due to the variability of scales, the present horizon R would only be half the size of the light travel time, i.e. about 6.8 billion light years.

In addition to the effect on wavelengths λ and frequencies f (which define the units meter and second), accelerations also appear smaller because they are measured in the variable unit m/s². Since the unit of force remains constant, Newton's second law

---

[1] This formula is about proportionality, so it can still contain numerical factors.

5 The Cosmos Without Expansion: A World of Variable Scales

$F=m \cdot a$ requires accelerations to be inversely proportional to masses. Therefore, in a universe where wavelengths and frequencies are decreasing over time, masses appear to be more inert and thus heavier. This will become important later.

GRAVE OVERSIGHT WITH SERIOUS CONSEQUENCES

Unfortunately, Dicke failed to complete the revolution and to confirm one of Dirac's hypotheses. For he believed that the expression for the number of particles $N \sim t^{\frac{3}{2}}$ (see table) would contradict Dirac's second conjecture, which had assumed a proportionality of $N \sim t^2$.

Since Dicke's argument for $N \sim t^{\frac{3}{2}}$ seemed quite contrived to me, having gone through the article several times I noticed that he had apparently confounded[24] the absolute time t with the observable time t', a quantity he himself had defined correctly and, due to the variable scales, must develop as $t^{-\frac{3}{4}}$.

However, Dirac's conjecture – with which I had been familiar for some time – is fulfilled[25] if one writes the particle number as a function of the *observed* time t': $N \sim t'^2$. Since the radius of the proton r_p develops in the same way as the wavelengths λ – as can be easily seen from the figures in the table – the coincidence

$$\frac{M_u}{m_p} \approx \frac{R_u^2}{r_p^2}$$

must hold true. This relation establishes the connection between cosmology and elementary particles and demonstrates that another constant of nature can be eliminated. However, since the numerical agreement is only approximate, the calculation of another pure number is necessary. If this succeeds – probably only after a deeper understanding of the proton and its 'radius' – then

Part II: The End of Space and Time

of the nine constants initially considered, G, h, c, M_u, m_p, R_u, r_p, 137 ..., 1836 ..., only seven would be independent.

It is useful to summarize this new picture of the evolution of the cosmos which continues the ideas of Mach, Einstein, Dicke and Dirac. First of all, the iconic feature of the conventional model, the Big Bang, needs to be completely reconceived: There is no material expansion of the cosmos. This impression is only created by the fact that atoms shrink with time and therefore reduce their wavelengths λ, providing smaller and smaller measuring rods but not greater distances. On the other hand, the universe is by no means static,[1] because the scales are continuously changing as light spreads. Now one may mentally go back in time to the moment when the propagation of light began. We should call this instant in the distant past the 'Big Flash', to emphasize that since then only light, but not matter, has been spreading.

A UNIVERSE OF CHANGE

For the reader unfamiliar with mathematical notation, the easiest way to understand the variable quantities is by looking up the example in the second column from the right in the above table. Imagine that the visible horizon of the universe R shortly after the primordial 'Big Flash' was just as big as one elementary particle (radius of the proton r_p). All physical quantities such as lengths, times etc. refer to this moment, i.e. t=1, f=1, R=1 etc. Consider then the situation after the (absolute) time step no. 10,000. Since the horizon grows only with the square root of the time t, it has increased to 100 times its initial value, while at the same time the speed of light has decreased to 1/100 of its original

---

[1] Not even stationary. In the 1960 and 70s, there was an intense debate between 'steady state' proponents and adherents of the Big Bang, both providing reasonable arguments against the respective adversarial model. Indeed, the model presented here suggests that both were wrong in a subtle manner.

5 The Cosmos Without Expansion: A World of Variable Scales

amount. Constrained by the equation $c=\lambda f$, this factor 1/100 must be evenly shared between wavelengths and frequencies, which accordingly drop to 1/10 of their original values. It follows, however, that with these 10-fold shortened scales λ, the 100-fold expansion of the universe appears 1,000-fold to an observer. Likewise, the decrease of frequencies leads to a 10-fold dilation of time scales that makes the $10,000^{th}$ time step appear only as the $1,000^{th}$. For an observer who is unaware of the variability of the speed of light, the cosmos appears to have expanded 1,000-fold during 1,000 time steps, i.e. constantly, conveying the illusion of an unchanging expansion rate c!

It is easy to see that the volume of a three-dimensional sphere in which light propagates has increased to one million times its original volume, i.e. by a factor of 10^6. If one assumes that the density, i.e. the number of particles per (absolute) volume in the universe is a constant – another simple assumption on which Dicke's visionary model is based – then after the $10,000^{th}$ time step, one million particles are also visible. However, this particle number 10^6 appears as the square of the visible time step $1,000=10^3$, if one takes into account the changing measuring scales. The above-mentioned example can now easily be extended by considering the time step 10^{52} instead of 10,000 (52 zeros instead of 4), see third column from the right. It now becomes obvious that this model exactly reproduces Dirac's observation. The fact that the number of particles is just proportional to the *square* of the length ratios in the universe necessarily follows from the idea of variable scales. Ultimately, therefore, Dirac's second observation can already be deduced from Einstein's idea of the variable speed of light from 1911, if one thinks a few steps ahead.

Part II: The End of Space and Time

HISTORICAL LONELINESS

It is a pity that the founding fathers of this scientific revolution could never get to exchange their best ideas. If they had, they would surely have recognized the similarities, and their ideas would be given the attention they deserve. Dirac knew nothing about Einstein's attempts (and vice versa), which were also unfamiliar to Dicke. Instead of joining forces with Dirac, the two visionaries got caught up in a petty argument,[26] because Dicke had not realized the close connection between his own theory and Dirac's second hypothesis.

And of course, it is amazing how little the current cosmological fashions deal with these fundamental reflections. On the one hand, there is certainly a gross historical ignorance among today's practitioners who do not care to read allegedly outdated publications from one century ago. On the other hand, that entire research fields can be locked up in a false paradigm that has become entrenched after decades of research tradition is not really a surprise to readers of Thomas Kuhn. Today, people no longer ask fundamental questions like Einstein and Dirac did, but are happy with an undemanding business of fitting arbitrary parameters. If one seeks a rational picture of nature, however, the goal must be to understand unexplained numbers, and the relation

$$\frac{M_u}{m_p} \approx \frac{R_u^2}{r_p^2}$$

in this respect, represents substantial progress that reduces the number of independent free parameters. Hence, as far as epistemology is concerned, calculating the gravitational constant G and deriving Dirac's second hypothesis are the most important result of the variable speed of light cosmology that goes back to Einstein and Dicke. Yet it does have further consequences.

5 The Cosmos Without Expansion: A World of Variable Scales

ENLIGHTENING DARK ENERGY

In the conventional view, the supposed expansion of the universe is slowed down by the effect of gravitation, and a deceleration of this expansion had therefore been expected. However, this deceleration was simply not observed.[27] To fix the problem, a quantity called 'dark energy' was invoked for the very purpose of compensating the supposed deceleration of the expansion with an ad hoc postulated 'accelerated expansion'. Instead of such a contrived construction of two mysterious antagonists neutralizing each other (which in addition, requires a bizarre fine tuning), a much simpler picture emerges: Since the expansion itself is an illusion, there is no reason whatsoever why it should be accelerated or decelerated. The present cosmological model merely describes what we observe: a seemingly constant expansion of the universe. Therefore, there is no more need for 'dark energy',[28] an unexplained numerical value that, due to its questionable origin, we had not even deemed a respectable constant of nature.

The adherents of the standard 'concordance model' of cosmology – which has grown steadily more complicated in recent decades and now contains as many as 17 parameters[1] – may not be put off for the time being. However, no matter how many alleged confirmations of that model are published, they cannot hide the methodological hamster wheel on which the research field finds itself. It is obvious that the conventional model describes numerous contradictions with more and more arbitrary numbers, but has failed to understand the very beginning of quantitative cosmology, namely Hubble's redshift. From the point of view of natural philosophy, to resort to such an unexplained expansion

---

[1] According to the renowned cosmologist Mike Disney, who, by contrast, counts only 13 independent measurements, "This situation is anything but healthy" (arXiv.org/abs/astro-ph/0009020).

Part II: The End of Space and Time

constituted the first step astray, and it was only Robert Dicke's insights that opened the doors to a real understanding.

Contrary to what Dirac was dealing with in 1938, however, this is not just an interesting speculation, but the consequence of a worked-out model that not only simplifies cosmology by explaining the redshift, but also incorporates general relativity, considered one of the two cornerstones of physics. As with all real progress in physics, the three elements of vision-mathematization-simplification can be recognized here, even if the origins go back to Ernst Mach in the 19th century and the results have not yet entered the scientific mainstream.

6 Revolutions That Have Not Yet Taken Place

The last two chapters were dedicated to two radical simplifications: the explanation of the gravitational constant G by the masses in the universe and the confirmation of one of Dirac's two hypotheses by the variable speed of light model. Accordingly, two important constants have been eliminated.

The key to these two findings was hidden in the strength of the gravitational constant $G \approx c^2 R_u/M_u$ and in Dirac's observation regarding the size and mass of particles and the universe: $\frac{M_u}{m_p} \approx \frac{R_u^2}{r_p^2}$. One may assume that such numerical coincidences continue to play a pioneering role.

Dirac's second hypothesis was justified by the mathematical model presented in the previous chapter, but what about the first hypothesis that related the ratio of the electric and gravitational forces to the size of the universe? This seems like a hopeless task, and you might guess that the riddle will not be solved unless we have a unified theory of electrodynamics and gravitation – that holy grail of physics in which generations of physicists have failed to make progress.

Surprisingly, however, it turns out that Dirac's first conjecture $F_e/F_g \approx R_u/r_p$ can be formulated in a completely different way that is not only much simpler and more intuitive, but also suggests that a theoretical framework is not completely out of reach.

AN IMPORTANT ASSUMPTION

Actually, it has been known for a long time[29] that Planck's constant h is approximately equal to the product of the speed of light, the mass of the proton and its radius:

Part II: The End of Space and Time

$$h \approx c\, m_p\, r_p.$$

Of course, this formula displays the definition of the Compton wavelength $\lambda_C = \frac{h}{cm_p}$, a well-known quantity in quantum mechanics. However, according to common wisdom, the wavelength λ_C calculated from the mass alone does not reflect the actual size of a particle. Rather, most physicists would argue that[30] the masses and, accordingly, the Compton wavelengths of elementary particles have nothing to do with their size and that the case of the proton is just a random coincidence. Accordingly, the proton is not given a prominent role among elementary particles. In reality, however, it is the only particle in the universe that is massive and stable at the same time. The fact that its Compton wavelength approximately matches its real extension measured by experiments is a clear indication of the paramount role of the proton in fundamental physics. I am fully aware that this is against the current fashion, so let me quote Einstein on the issue, who was also convinced that the size of elementary particles had a meaning:[31]

> "The real laws are much more restrictive than those we know. For example, it would not violate the known laws if we found electrons of any size (...). Nature however realizes only electrons of a certain size (...)."

A couple of years ago, the charge radius of the proton was determined to be somewhat smaller than previously believed,[32] namely $r_p=0.841\cdot 10^{-15}$ m. It is remarkable that the approximate agreement in the above formula is significantly improved by a simple factor:[1] $h=\pi/2\; c\, m_p\, r_p$. The formula is even valid within the current measuring limits of about one percent!

---

[1] Even I do not endorse those approaches; the factor $\pi/2$ has already been noted by several authors, e.g. Dirk Freyling's 'mass-radius constancy' (www.ek-

Since this formula contains fundamental constants of nature only, it would be extremely important to derive it from a theory. Although I would like nothing more than to show you such a theory, it probably does not exist yet, and its development will certainly not be child's play. However, there is the more general question of what progress physics can hope for at all within the current system of constants of nature. I shall therefore assume in the following that the development of such a theory is possible as a matter of principle. In any case, there are no compelling reasons why it should not be.

Incidentally, there is a simple reason why, *without* Dirac's first conjecture, there can be no further progress at all in understanding elementary particles. A thorough understanding would require a calculation of their masses, which is literally unthinkable in the current paradigm, because the constants h, c, e, ε_0 etc. cannot be combined in a way that the unit of a mass, kg, emerges[1]. Such a calculation would be feasible only if the gravitational constant G is included. However, then the large numbers observed by Dirac would automatically appear, a consequence of the fact that the very nature of mass can only be understood cosmologically, as Ernst Mach had suspected. This simple argument is nevertheless widely unknown.

DIRAC EVERYWHERE

However, let us now illustrate the connection to the bold conjecture of Paul Dirac in 1937. Since the inverse-square law dependence of r cancels out, the ratio of electric force and gravitational force in the hydrogen atom is given by:

---

theory.com/) and N. Haramein (resonance.is/wp-content/uploads/QGHM.pdf)
[1] A formal proof can be found in A. Unzicker, www.arxiv.org/abs/9612061, sec. 4. One could rephrase the problem by stating that in order to calculate the m_p, r_p is necessary. Obviously, the radius of the proton is a relevant quantity.

Part II: The End of Space and Time

$$\frac{F_e}{F_g} = \frac{e^2}{4\pi\varepsilon_0 G m_p m_e}.$$

Now we can use the formula for the gravitational constant $G = \frac{c^2 R_u}{6 M_u}$ developed in Chapter 4 and obtain:

$$\frac{F_e}{F_g} = \frac{6 M_u e^2}{4\pi\varepsilon_0 c^2 R_u m_p m_e}.$$

After inserting the definition of the already mentioned fine structure constant $\alpha = \frac{e^2}{2hc\varepsilon_0} \approx \frac{1}{137}$ and the mass ratio between proton and electron $m_p/m_e = 1836.15...$, we get

$$\frac{F_e}{F_g} = \frac{6 \cdot 1836 \, M_u h c}{2 \cdot 137 \, \pi c^2 R_u m_p^2}.$$

In this seemingly complicated expression, we can substitute the quantum of action with $h = \pi/2 \, c \, m_p \, r_p$, which yields

$$\frac{F_e}{F_g} = \frac{6 \cdot 1836 \, M_u c^2 \pi m_p r_p}{137 \cdot 4\pi c^2 R_u m_p^2},$$

or, after cancelling various quantities,

$$\frac{F_e}{F_g} = \frac{3 \cdot 1836 \, M_u \, r_p}{2 \cdot 137 \, R_u m_p}.$$

Except pure numbers, this contains only those quantities that already occur in Dirac's second conjecture, vindicated in the previous chapter. If one inserts $\frac{M_u}{m_p} = \frac{R_u^2}{r_p^2}$ and neglects pure numbers, it becomes immediately clear that the first conjecture formulated by Dirac

$$\frac{F_e}{F_g} \approx \frac{R_u}{r_p} = \tau \, (\text{,epoch'})$$

holds true within orders of magnitude.[33] $h = \pi/2 \, c \, m_p \, r_p$ is thus practically equivalent to Dirac's first hypothesis! The fact that we have made use of $1/\alpha \approx 137$ means, of course, that the link from

$h = \pi/2 \; c \; m_p \; r_p$ to the electro-gravitational force ratio F_e/F_g will only be completely established when the number 137 has been calculated. This seems to be anything but an easy endeavor, since the problem has defied the attempts of generations of theoreticians. The same can be said of the number 1836, the proton-to-electron mass ratio.

Paul Adrien Maurice Dirac (1902-1984)

Since Paul Dirac, legions of physicists have been pondering over the origin of this number, with the same lack of success. However, as α=1/137.035999, it is 'just' a pure number, and again there is no theoretical obstacle that would inhibit a genius from calculating the m_p/m_e mass ratio.

Part II: The End of Space and Time

There are still other reasons why it makes sense to rewrite Dirac's first hypothesis in the form $h = \pi/2 \, c \, m_p \, r_p$. The similarity with the formula $\hbar = v \, m_e \, r_B$ leaps to the eye. Nils Bohr had used it in 1913 and triggered the revolution in atomic physics by giving h the role of an angular momentum of an electron. Since we have the velocity c here, the obvious guess is to see h as the angular momentum of a circular light wave that could represent a proton. However, a consistent model with some chance of being generally accepted has not yet been developed.[34] The difficulties seem enormous, not least because the reason the proton is carrying the elementary charge would have to be explained at the same time. If this were to succeed, it would be nothing less than a unification of quantum theory and relativity, whose constants h and c appear in the formula.

In his 1922 doctoral thesis, the French physicist Louis Victor de Broglie made an interesting attempt (though not specifically addressing the proton) towards this goal by amalgamating two of Einstein's famous formulae, namely $E = hf$ and $E = mc^2$, into the equation

$$hf = mc^2.$$

He argued that if one wants to bring the wave and particle aspect into a unified picture, equating the two energy terms would be the first step to take.[35] Intriguingly, when multiplying $h = \pi/2 \, c \, m_p \, r_p$ by f and considering the circumference $2 \pi r_p$ of a circular light wave, one obtains a very similar expression

$$hf = \tfrac{1}{4} m_p c^2,$$

which differs from de Broglie's formula only by a factor. This reminds us of yet another context. As pointed out in Chapter 4, $\tfrac{1}{4} c^2$ acts as gravitational potential, if one formulates the theory of general relativity by means of variable speed of light. It is cer-

6 Revolutions That Have Not Yet Taken Place

tainly not easy to make sense out of all these coincidences. However, it is remarkable how similar the grounds were in which Bohr, de Broglie, and Dirac were hunting for a unified theory.

AGE OF THE COSMOS – A FUNDAMENTAL CONSTANT?

Another coincidence that might stimulate further ideas should be mentioned here, although it does not add anything fundamentally new. As is well known, Planck's quantum of action h has the dimension of angular momentum, $kg\, m^2/s$, and it is an obvious idea to compare this smallest possible angular momentum with the largest possible one in the universe.

Whatever astrophysical object we may identify at the horizon, the maximum speed it can rotate around us is c, which limits the total angular momentum to $L = c\, M_u\, R_u \approx 10^{117}\, \hbar$, a huge number that contains, for obvious reasons, the epoch to the third power. Since this equation contains the above $h = \pi/2\, c\, m_p\, r_p$ and the already proven $\frac{M_u}{m_p} \approx \frac{R_u^2}{r_p^2}$, it may be seen as a direct consequence of Dirac's coincidences.

With regard to the Newton bucket experiment discussed in Chapter 4, however, we should remember Ernst Mach's spirited objection that a rotation of the universe would probably be unobservable in principle. For we define an (unaccelerated) rest frame precisely as a coordinate system in which the universe does not rotate. Due to the maximum speed c and the existence of the horizon, however, there are limits to that rotation rate anyway, namely $2\pi\, R_u/c$. Comparing such a rotation with velocity c at the horizon to the minimum angular momentum $\hbar = h/2\pi$ appearing in nature could be a starting point for exploring the origin of the constant h, especially at the time of the 'Big Flash' at t=1.

Part II: The End of Space and Time

If one imagines a rotation of the universe with velocity c, it is interesting to consider the centripetal acceleration $a_z = c^2/R_u$. In recent decades, there have been a considerable number of surprising observations that have usually been interpreted as 'dark matter'. Practically all these observations however occur in a small acceleration regime[36] around and below c^2/R_u.[I] This, together with the fact that dark matter requires several free parameters (e.g. for its distribution) is highly suspicious and indicates that these anomalies are due to a still insufficient understanding of gravitation. The magnitude c^2/R_u is indeed a clue that points to a cosmological origin.[II]

WHY THE PLANCK UNITS ARE NOT HELPFUL

There is a vast literature on the coincidences found by Dirac, many of them published in variant forms,[III] and some of which are so different that the authors were unaware of having rediscovered Dirac's findings. In fact, little news is expected from papers claiming to have found new relations between fundamental constants. What has been known for a long time, however, is that the units m, s and kg may be derived from G, h and c.

The quantities $l_{pl} = \sqrt{\frac{Gh}{c^3}}$, $t_{pl} = \sqrt{\frac{Gh}{c^5}}$, $m_{pl} = \sqrt{\frac{hc}{G}}$ are named after Max Planck as Planck length, Planck time and Planck mass. Due to the lack of reference to cosmology, which did not exist at the

---

[I] This holds particularly for the edges of galaxies, but is also observed, for example, in globular clusters. A comprehensive account of these phenomena can be found in the book *The Dark Matter Problem* by Robert Sanders (2010).

[II] The British physicist Mike McCulloch also considers the relationship between masses and acceleration in his theory of *quantized inertia,* which discusses the well-known Casimir effect on a cosmological scale. Here, too, the small acceleration c/t_u, occurs, just like in the alternative gravitational theory MOND (*MOdified Newtonian Dynamics*).

[III] Some include Mach's principle $c^2 R_u = G M_u$, which allows for another variety of possible expressions.

time, these expressions published in 1910 provide very limited insight. In contrast to the experimentally accessible quantities m_p and r_p, there is not the remotest chance of testing the validity of the Planck units l_{pl} and r_{pl}. If the relations $h = \pi/2\ c\ m_p\ r_p$ and $G = \frac{c^2 R_u}{6 M_u}$ are inserted into the Planck units, it turns out that r_p and m_p differ by a factor of 10^{20}, i.e. the square root of the epoch. Although the Planck units are sometimes called 'fundamental', most people overlook this relation to Dirac, which is their only nontrivial property.

A prominent example of someone who believed he had found relations among fundamental constants was Carl Friedrich von Weizsäcker, a student of Heisenberg. He considered the approximate coincidence

$$m_p^3 \approx \frac{h^2}{G\ R_u},$$

which was recently improved by the Viennese engineer Helmut Söllinger to

$$\sqrt{m_p m_e} = \sqrt[3]{\frac{e^2 h}{4\pi \varepsilon_0 c G R_u}},$$

thereby obtaining a much better accuracy. Yet, the expressions can be traced back to Dirac's conjectures by using Mach's principle $c^2\ R_u = G\ M_u$, the definition of the fine structure constant $1/137 = ...$, and the formula $h = \pi/2\ c\ m_p\ r_p$. Only a proper theory that justifies the latter ones will lead to significant progress, however. Intriguingly, the potential of Dirac's conjectures is indicated by another, albeit approximate, agreement first mentioned by George Gamow.[37]

Since the number 137 is a bit too large to be easily generated by, say, combinations of π and small integers, the odds of a theoretical breakthrough seem to be low. However, in math and many applied sciences the logarithm is known as a function that transforms very large numbers into small, 'handy' ones. Since $\log(\tau \approx 10^{40}) \approx 92$, it is not too far off to speculate about the natural logarithm of the epoch τ being related to the fine structure constant $\alpha = \frac{1}{137}$, e.g. by assuming $\frac{3}{2\alpha} = \log \tau$.

In calculus, the log function occasionally appears when integrating the function 1/r, which is, after all, not unlikely to occur in a cosmological context – remember the expression for the gravitational constant G developed in Chapter 4. Of course, Gamov's speculative idea still needs a thorough theoretical justification. Even more hypothetical would be assuming a possible link between $\log \tau$ and the number 1836, for which there is not even a conceptual idea so far. However, since the order of magnitude does not differ that much from 137, one could hope that a breakthrough in understanding the fine structure constant would also shed light on the mass ratio m_p/m_e of proton to electron.

From a general perspective, the question is to what extent the number of constants of nature can still be reduced. As history shows, each individual constant has created and will create its own peculiar, sometimes tremendous difficulties. Yet it is not elementary logic or some other basic law that prohibits the solution of such riddles as a matter of principle. Hence, to assume that these theoretical breakthroughs will ultimately be made – for example by calculating the fine structure constant – is certainly optimistic. On the other hand, there is no other consistent perspective if we do not accept God-given constants, but consider them as an incentive for future efforts to understand nature.

6 Revolutions That Have Not Yet Taken Place

DOES A NEUTRON FEEL THE AGE OF THE UNIVERSE?

A final coincidence, related to the hitherto neglected neutron, was formulated by Pascal Jordan who was involved in two Nobel-worthy discoveries[I]. From a philosophical point of view, the half-life of the neutron t_n of about ten minutes is a basic quantity that calls for an explanation. If we take a look at the scheme of variable scales in Chapter 5, we notice that t_n is of same order of magnitude as the square root of the absolute time step $t=10^{53}$, amounting to $10^{26,5}\cdot r_p/c \approx 900s$. Claiming that the decay of the neutron can be deduced within the variable speed of light model would certainly be premature. In any case, however, the phenomenon of radioactivity will only be thoroughly understood once t_n is calculated from first principles. Only then can we hope to compute the small mass difference between proton and neutron and – another significant puzzle – understand why[II] some mass seems to be lost when the neutron is converted into a proton.

However, the proof of Dirac's coincidence $\frac{M_u}{m_p} \approx \frac{R_u^2}{r_p^2}$ given above motivates us to look for a connection between the proper-

---

[I] Jordan had given his mentor Max Born a draft of a manuscript that Born had forgotten in his suitcase for several months. As a result, the priority in the publication was lost for what is called Fermi-Dirac statistics today. Likewise, Jordan significantly contributed to the statistical interpretation of the wave function commonly attributed to Max Born.

[II] In this so-called beta decay, a neutron transforms into a proton, an electron and an (anti-)neutrino, according to conventional wisdom. I cannot address the intricacies of the beta decay and the decade-lasting inconsistencies in neutrino physics here, but it is obvious that the model has grown to a suspicious complexity. My guess is that there is a fatal misconception behind the paradoxes of the entire neutrino, and Niels Bohr was presumably right. At a conference in Rome in 1930, he expressed doubt about energy conservation and suspected that physics had to await another revolution similar to atomic physics before the nucleus could be understood.

ties of the universe and those of elementary particles. The equation clearly shows that the size and mass of the proton are highly significant for a basic understanding. According to the model of changing measuring scales developed in Chapter 5, the visible horizon was just as big as an elementary particle at time t=1, while the universe back then was densely packed with particles. Light then spread at a much greater velocity than today, but at the same time atomic nuclei continued to shrink. Gradually, they occupied a smaller share of the volume of the universe, which today has dropped to the tiny fraction 10^{-40}. If one extrapolates back to the very first moment t=1, however, there are some interesting consequences. First it is rather hard to believe that in such a simple universe, a sophisticated physical theory applies that provides a number such as 1836. However, if the mass ratio m_p/m_e=1836 does depend logarithmically on the age of the universe, then it follows that at the time of the 'Big Flash' the electron and proton were of equal weight.

BIG SIMPLICITY AT THE BIG FLASH

The hydrogen atom would then be similar to an object now called positronium, consisting of an electron and its antiparticle positron that orbit each other. The definition of the fine structure constant implies that $1/\alpha \approx 137$ is the ratio of speed of light c to the electron's velocity on the innermost orbit of the hydrogen atom.

If, analogous to 1836, the number 137 depends in a logarithmic manner on the age of the universe, it would have been of the order 1 at the very first instant of the universe. This would imply that the orbital speed of the electron in the hydrogen atom was equal to the speed of light. This, in turn, suggests that the hydrogen atom – at that time an orbiting electron-positron pair [1] – could

---

[1] Pair generation, in which electron-positron pairs may be created from pure

6 Revolutions That Have Not Yet Taken Place

simply be seen as a rotating light wave. Even in the conventional view, the electromagnetic fields defining light contain virtual electron-positron pairs. Considering again α and the electron's velocity $v = \frac{Ze^2}{2h\varepsilon_0}$ proportional to the charge Z of the nucleus, 137 obviously represents an upper limit for the number of protons contained in an atom; otherwise the electron speed would exceed c.[I] So if today's 137 had the value 1 at the 'Big Flash', atoms heavier than hydrogen could not exist, apart from the fact that the astrophysical conditions for their formation occurred much later.[II] If one adds the hypothesis about the cosmological origin of the half-life of the neutron, initially t_n would have been infinitely small, implying that the neutron could not have existed in a stable manner. Again, it would be equivalent to a primordial hydrogen atom/positronium.

In summary, if one dares to extrapolate back to the 'Big Flash' while systematically applying the system of variable scales outlined in Chapter 5, a picture emerges in which one can regard the neutron, the hydrogen atom, an electron-positron pair and a circular light wave as the very same object. This is certainly not unappealing. At the same time, the expansion of the universe would gradually create the conditions for heavier nuclei and thus for all other evolutionary processes in the cosmos.

Such a scenario is certainly speculative, and the coincidences pointing in its favor are hardly more than a clue to an exciting research topic. Given that some 'constants' of nature have proven to be variable, it is only natural to question the dogma of the

---

light, as well as the reverse process of pair annihilation, are interesting phenomena. There is, however, no justification for their existence being derived from first principles.
[I] This holds true even if there is a relativistic increase in the electron mass.
[II] It is generally believed that heavy nuclei can only be produced during supernova explosions, thus obviously requiring the formation of stars.

Part II: The End of Space and Time

unchangeability of the laws of nature. Dirac had argued the following in 1968:

> *"Theoretical workers have been busy constructing various models for the universe based on any assumptions that they fancy. These models are probably all wrong. It is usually assumed that the laws of nature have always been the same as they are now. There is no justification for this. The laws may be changing, and in particular, quantities which are considered to be constants of nature may be varying with cosmological time. Such variations would completely upset the model makers."*

In any case, it remains the task of theoretical physics to explore the origin of seemingly arbitrary quantities occurring in nature. It is quite obvious that these fundamental constants are in one way or another related to the evolution of the cosmos.

7 The Origin of Mass and the Riddle of Physical Units

If we now count the constants of nature that are left over from the nine constants G, h, c, M_u, R_u, m_p, r_p, 137 and 1836 considered initially, the following picture emerges: the expression for the gravitational constant G outlined in Chapter 4 and the deduction of Dirac's second coincidence $\frac{M_u}{m_p} \approx \frac{R_u^2}{r_p^2}$ reduces the number by two, leaving only seven independent constants.[1] If the reducing equation $h = \pi/2\, c\, m_p\, r_p$ is proven to be correct, six constants remain, and under the additional assumption that the two numbers 137 and 1836 can be calculated, only four.

It turns out that there is another constant that can be eliminated, but first we need to reflect on the system of physical units. Physical reality is based on the concepts of space, time, and mass, with their respective units of meter, second, and kilogram. These units are inseparably interwoven with the existing constants of nature, because all lengths, times, and masses can be written as multiples of the Planck units. Given that the gravitational constant G could be calculated from the mass distribution of the universe and has thus become a superfluous quantity, the question arises as to whether one physical unit, conveniently the kilogram, might also be obsolete.

UNITS AND CONSTANTS OF NATURE

The elimination of the kilogram would require a more thorough understanding of the nature of mass. In Einstein's principle

---

[1] Strictly speaking, there is still a numerical value to be calculated, which turns Dirac's approximate coincidence into a precise one.

Part II: The End of Space and Time

of equivalence, inertial mass is characterized by resisting acceleration. Accordingly, in Newton's second law, $F = m \cdot a$, mass is proportional to inverse acceleration that bears the units s^2/m. Inverse acceleration is the very quantity by which the mass ratios of celestial bodies have been determined since Kepler's time. In this respect, redefining mass in terms of inverse accelerations would be the seamless incorporation of Newton's second law into physics.[I] However, a complete understanding would also have to reflect Ernst Mach's conjecture that the origin of inertia was the existence of all other masses in the universe. Mach had also suggested defining mass by inverse acceleration. Such a theory would have to convert the existing kilogram unit into s^2/m by determining a numerical factor, and thus clarify the origin of inertia quantitatively.[II]

Unfortunately, is completely obscure what such a theory of inertia should look like, and I do not foresee an opportunity to develop it before the modified Dirac conjecture $h = \pi/2 \ c \ m_p \ r_p$ is justified; there is a hint, however. If one assumes that the proton is rotating in some manner (as the angular momentum suggests), it must be subject to a centripetal acceleration of the order of c^2/r_p, the inverse of which could be equated with the proton mass m_p. Therefore we get

$$m_p \ c^2/r_p = 1 \text{ or } 1 \text{ kg} = 5{,}8 \cdot 10^{-5} \ s^2/m.$$

While attempting to solve the problem in the microscopic realm, it is imperative to take Mach's principle into account, which is realized by the expression $\frac{1}{c^2} = \frac{\kappa}{2\pi} \sum \frac{m_i}{r_i}$ discussed in Chapter 4. This means that the constant $\kappa \approx 10^{-37}$ would have to

---

[I] It should be noted, however, that $F = m \cdot a$ does not apply to strongly accelerated charges, since they radiate part of the acceleration work. See Landau-Lifschitz II, Chapter 75.

[II] The idea of choosing inverse acceleration as the unit of mass has recently been taken up by the British physicist Julian Barbour.

7 The Origin of Mass and the Riddle of Physical Units

assume a pure numerical value, which corresponds approximately to the reciprocal value of the epoch τ.[1]

I am aware of piling optimistic assumptions on top of each other, yet there seems to be no other way to get rid of the 'gods of modernity'. In any case, if a theory is developed that deduces $m_p = \frac{r_p}{c^2}$, one may insert it into the formula $h = \pi/2 \, c \, m_p \, r_p$ (which is still to be established), obtaining

$$hc = \pi \, r_p^2.$$

This connects the fundamental constants h and c with the last unexplained quantity r_p, whereby the product hc may be interpreted as the 'area of the proton'. Planck's constant h would thus have the value $3{,}7 \cdot 10^{-39}$ ms – i.e. a unit consisting of the product of space and time. If this proves to be correct, then a total of four reducing equations (the above-mentioned one, as well as those explained in Chapters 4, 5, and 6) would reduce the number of constants of nature from seven (G, h, c, Mu, Ru, m_p, r_p) to three (h, c, and the epoch $\tau = R_u/r_p$).[II]

RAISING THE INVENTORY OF SIMPLICITY

More simplification seems to be impossible at the moment, however optimistic one might be about the ingenuity of theoreticians. Before we explore the deeper reasons for this, it is worthwhile recapitulating the results so far. A maximum of conceptual simplicity had been achieved by around 1930, and this was the moment we turned away from the prevailing opinion and followed a different epistemological path. Which unifications had

---

[1] This is not completely satisfactory, since the time dependency should already be included in the sum. However, setting $\kappa=1$ for the mass definition yields nonsensical values; thus it is more likely that the problem is solved by taking into account the evolution of scales described in Chap. 5.

[II] It has already been assumed that the pure numbers 137 … and 1836… can be computed theoretically.

Part II: The End of Space and Time

succeeded and which might possibly come into sight? In any case, it turned out that scientific revolutions followed the pattern of vision-mathematization-simplification, leading to an elimination of constants of nature.

If we consider the various fields of physics – such as cosmology, gravitation, mechanics, optics, thermodynamics, electrodynamics, quantum mechanics, and nuclear physics – many cross-connections have been established by explaining fundamental constants. Newton, the founding father of modern science, succeeded in unifying earthly mechanics with the gravitation of the solar system. Much later, in 1905, Einstein established a connection between mechanics and optics by realizing the relevance of the speed of light c for dynamics. General relativity can be seen as a unification of gravitation and optics, once we consider its natural form from 1911 that was much later developed by Robert Dicke. By relying on ideas of Ernst Mach, Dicke thus extended a bridge from gravity to cosmology, which had been designed by Dirac in 1938. At the same time, Dirac had connected cosmology with atomic physics, although this unification has not yet been fully worked out.

Thermodynamics was unified with mechanics at an early stage by the findings of Robert Mayer and James Prescott Joule, and completed in 1906 by Ludwig Boltzmann. Max Planck combined thermodynamics and optics in his radiation law, which was the first hint of the existence of quantum mechanics. After all, it is due to Maxwell, Weber,[1] and Hertz that optics can be regarded as being unified with electrodynamics.

---

[1] Weber also developed a theory of electrodynamics that differs from Maxwell's. See Assis (1994).

7 The Origin of Mass and the Riddle of Physical Units

THE FAKE MARRIAGES OF PHYSICS

However, there is no real unification of electrodynamics and quantum mechanics, even though a theory called quantum electrodynamics is considered successful. Its predictions are based on interesting, but not really fundamental properties of nature.[1] Often, due to hidden assumptions, an exaggerated precision of measurements is reported. Fatal internal contradictions of the theory have never been clarified – a fact, incidentally, even conceded by Richard Feynman (he received a Nobel Prize for the theory).[38] A reasonable definition of 'quantum electrodynamics' deserving of the name would be a theory that actually combines the constants of electrodynamics, namely the elementary charge e and the electric field constant ε_0, with the constant of quantum theory h. However, that means justifying the equation

$$\frac{e^2}{2hc\varepsilon_0} \approx \frac{1}{137},$$

or, as already mentioned, calculating the fine structure constant. This would also require a conceptual clarification of how the quantization of h is related to the quantization of the elementary charge e. A derivation of the number $m_p/m_e \approx 1836$ could finally be seen as a unification of atomic and nuclear physics, which will probably only be possible after the complete realization of Paul Dirac's dream of connecting cosmology with nuclear physics.

> *Quantum electrodynamics is a complete departure from logic. It changes the entire character of the theory.* – Paul Dirac

---

[1] Usually the Lamb Shift and the anomalous magnetic moment of the electron are invoked here, which can be calculated with the help of quantum electrodynamics. But the question arises why a really fundamental theory is unable to compute the much more important number 137.

Part II: The End of Space and Time

QUANTUM GRAVITY BEGINS IN THE PROTON

Quantum theory has remained an alien component of physics until today. This is even the conventional view that emphasizes the incompatibility with the theory of general relativity. It is usually mathematical formalism that is seen as the culprit. Quantum mechanics, as we are told, 'lives' in a normal Euclidean space, while a curved space-time is believed to be the mathematical niche general relativity needs to flourish.

It is true that these incommensurate formalisms are an obvious obstacle for a possible unification of the theories. However, since the apparent 'curvature' of a space is mathematically equivalent to a variable speed of light, the formulation based on Einstein's 1911 idea, the latter approach is certainly more promising than a 'geometric' space-time for this reason alone. Although this is a relatively simple argument, it has nevertheless not become widely known among today's physicists.

This is not the place to give an overview of the failed ideas on 'quantum gravity', for decades an almost ubiquitous term in physics journals. It is tempting to assume that when so many people write about 'quantum gravity,' they must know what they are writing about.[39] But any theory worthy of the name would have to be able to calculate the balance of forces in the most basic quantum system – i.e. in a hydrogen atom – in other words, the number

$$\frac{F_e}{F_g} = 2{,}3 \cdot 10^{39}.$$

The only quantitative idea in this direction goes back, as mentioned, to Paul Dirac, and in the previous chapter it was shown that this number can be calculated in principle, once the relationship $h = \pi/2 \, c \, m_p \, r_p$ is justified. Trying to derive it would therefore be the only reasonable path towards 'quantum gravity'. For Planck's quantum of action h=6.626·10⁻³⁴ kg m²/s carries the

7 The Origin of Mass and the Riddle of Physical Units

whole mystery of quantum theory within itself. A consistent mathematical apparatus justifying the above equation would probably come very close to the desired unification of gravity and quantum theory.

UNANSWERED BASIC QUESTIONS

Less than mathematical formalisms, it is in first place the conceptual hurdles that must be overcome when unifying theories. By this I mean characteristics of nature that cannot be comprehended with pure logic, i.e. everything that makes you ask: Why does it work this way and not another way? The numbers 137 and 1836 are obvious examples, but there are still more basic riddles. Why is gravitation only attractive, but the electromagnetic interaction both attractive and repulsive? It may be that the solution is delivered when calculating the fine structure constant, but perhaps this key question must be answered first.

For example, the question 'Why does the universe seem to be expanding?' was answered in Chapter 6, but many others remain in the dark. What is the cause of radioactivity? Is a physics without radioactivity even conceivable, and if not, why not? Why are there two types of elementary particles, 'fermions' with half integer spin and 'bosons' with integer spin? Why does spin, the mysterious property of elementary particles that is linked to the constant h, exist at all? The latter question will be of particular interest for us.

Why is there no picture of quantum physics that is intuitively accessible? In some experiments, light seems to behave like a wave, in others like a particle. The same applies to all material 'particles': they, too, have been proved to display a wave nature. Most physicists have become accustomed to these paradoxes, but the terms 'wave-particle dualism' or the 'principle of complementarity' postulated by Niels Bohr,[1] are fig leaves rather than a real explanation.

GOOD QUESTION, NEXT QUESTION

Again, it is impossible to give even a sketchy overview of the variety of notions and concepts used to describe the intricacies of quantum theory. When prolonged efforts produce so little result, however, one must suspect that the problem has been tackled in a wrong way. More promising than discussing the nature of waves and particles is probably the question: Why does nature manifest itself in such characteristic yet distinct phenomena – light and matter?

This seems to be the real mystery, also because there is a parallel to the constants of nature, which have always provided clues to fundamental insights. Light is obviously associated with the speed of light c, while the constant h is a property of matter. Einstein's formula for the energy of light quanta $E = hf$ is not a counter-argument, because all experiments involving that formula (photoelectric effect, Compton effect) require the presence of matter. Whether $E=hf$ holds in a world without matter will therefore never be known. Equally hard to imagine is a 'dark' universe consisting of matter only (at zero temperature!) without any light. However, the question remains: Why did nature choose the two phenomena of light and matter? It seems obvious to relate this fact to the existence of the two constants of nature c and h, the only constants for which we could not even imagine what a possible calculation might look like.

---

[1] Bohr was a great physics visionary, but a considerable part of his writings were also notorious for their ambiguity and lengthiness.

7 The Origin of Mass and the Riddle of Physical Units

SIMPLICITY ALSO APPLIES TO THE UNITS

On the other hand, there is no reason why such an explanation that the human mind can understand should be precluded forever. If we accept the existence of the phenomena c and h without demur, they would indeed take on the role of the last gods of mankind.

A fundamental obstacle that hinders an explanation of h and c seems to be the physical units. As discussed above, eliminating the gravitational constant G (Chapter 4) by establishing the relationship $m_p = \frac{r_p}{c^2}$ would imply that the unit of Planck's constant h becomes m·s, the product of length and time. Just as the three constants of nature G, h, and c were used to define the quantities kilogram, meter and second via the Planck units (though these have no fundamental meaning), two constants of nature, h and c, are still needed to express the physical units meter and second, which is achieved in the expressions $\sqrt{hc}$ (meters) and $\sqrt{\frac{h}{c}}$ (second).

Hence, going even further and trying to eliminate the constants h and c, one would lose the units meter and second, the basis for any measurement in our physical reality. Obviously, the existence of h and c touches the foundations of the stage on which physics takes place: Space and time.

Thus, as soon as we no longer want to accept h and c as God-given entities, we must find a priori reasons for the existence of these elementary phenomena. Why do they show up at all within the framework of space and time? A related peculiarity, of course, is that space apparently has three dimensions and time only one: Why does nature present itself in this peculiar 3+1-dimensional fashion?

Part II: The End of Space and Time

DEAD END SPACE-TIME

This question has long been suppressed in an astonishing way, although by looking at history one may gain a better understanding of this collective blindness. After groundbreaking work by Hendrik Antoon Lorentz and Henri Poincaré, Einstein formulated the theory of special relativity in 1905. Space and time scales proved to be dependent on the motion of the observer, and were thus called 'relative'. All this was a consequence of the fact that the laws of nature themselves did not depend on the motion of the reference system.

An important mathematical tool for dealing with a moving reference frame is the so-called Lorentz transformation. Quite aesthetically, it establishes an analogy between ordinary spatial rotation (e.g. in the x-y plane) and motion (e.g. in the x-direction). The latter is mathematically described by a rotation in the x-t-plane, i.e. involving a spatial and a temporal dimension. When doing calculations, functions analogous to the well-known trigonometric quantities sine and cosine appear, which makes the formalism rather elegant. However, this does not prove there is a fundamental meaning behind it, since beauty is not a good measure for the relevance of a physical theory. For mathematicians who routinely deal with more than the three perceivable dimensions, however, it was tempting to invent the term 'space-time', conflating three-dimensional space with one-dimensional time to a four-dimensional structure.

Although the first idea for this came from Henri Poincaré, it was the German mathematician Hermann Minkowski in particular who stood out as an enthusiastic advocate of this concept. At the 1908 German congress of natural scientists in Cologne he gave a lecture with the legendary title "Space and Time". He captivated his audience with phrases such as the following:

7 The Origin of Mass and the Riddle of Physical Units

> *"Gentlemen! The views on space and time that I shall like to present are based on experimental physics and this is their strength. These views have a radical tendency. Henceforth space by itself, and time by itself, are doomed to fade away into mere shadows, and only a kind of union of the two will preserve an independent reality."*

Not that there wasn't any substance to that. Nevertheless, Minkowski's lecture had an enormous impact, and in retrospect one may see it as yet another proof of how psychology and sociology influence opinions in physics, the cherished queen of rational science. Even Einstein gradually gave up his initial wariness about the four-dimensional formulation. This, too, probably contributed to the fact that he later abandoned the variable speed of light formulation of general relativity in favor of geometric terms. Obviously, in such a variable speed of light theory, c would have played a central role encouraging further scrutiny. However, from Minkowski's point of view c became an unimportant 'conversion factor' whose sole purpose was to fix the different units of space and time.

SPACE AND TIME ARE NOT THE SAME

The four-dimensional picture, rationally speaking, is a denial of the reality established by 'modern' physics: Space and time are, as perceived by everyone who is not literally out of his senses, different phenomena of nature. We can navigate in space, but not in time, and not even stop it. Nevertheless, for more than a century theoretical physics, while considering itself an empirical science, has been pretending that the obvious difference between space and time does not exist. Since the time of Minkowski, doubts about the concept of 'space-time' have been so thoroughly erased from the collective memory that the vast majority of physicists would refuse to even acknowledge the problem.

Part II: The End of Space and Time

The consequences of this paradigm founded by Minkowski can only be called devastating. The development of special relativity and the emergence of the new constant h at the beginning of the 20th century could have highlighted the fundamental problem of the existence of space and time; yet it was completely ignored and swept under the rug by the concept of an allegedly four-dimensional 'space-time'.

Just as painting over rust is more harmful than visible corrosion, theoretical physics has suffered tremendously while carrying this misconception around for decades. Not coincidentally, the disregard for constants of nature and the ensuing pseudo-explanations have led to mathematical excesses such as string theory, which unfortunately ties up valuable resources among mathematicians. A few minutes of sober reflection instead would be enough to convince oneself that physics has not yet clarified the difference between space and time.

For the first time in this book, we now realize that a satisfactory theory of reality must question the origin of space and time and explore alternatives. Before doing so and thinking about how the concepts of space and time might be replaced, I shall remind you that there are still many properties of the constants c and h that are worth pondering over. Without a historical overview of these riddles it would be difficult to grasp the problem in its entirety. I shall therefore dedicate a chapter to each constant, in which you can familiarize yourself with the discoveries of h and c and the conceptual problems that arose during the course of history.

8 Finite Speed of Light: The Subtle Anomaly

It is alleged that even Galileo Galilei, probably the most diligent observer of nature of his time, speculated about the finite speed of light. Presumably, he had interpreted the high-frequency discharges during thunderstorms as reflections of lightning from distant clouds. Therefore, he concluded that the speed of light cannot be infinite – as it was considered an immutable truth by Aristotelian physics and its main representative, René Descartes. In any case, Galilei's legendary observation of the four moons of Jupiter in 1610 provided the basis for the first determination of the speed of light.

For a long time, the most precise data on Jupiter's moons had been gathered at the astronomical observatory in Copenhagen. In 1671, the French astronomer Jean Picard decided to go there in order to determine the exact difference in longitude between Copenhagen and Paris. Picard was so skillfully assisted by the Danish astronomer Ole Rømer that the former invited him the following year to visit the observatory in Paris, which was headed by the famous astronomer Giovanni Domenico Cassini.

THE HEAVENLY CLOCKWORKS

Their joint observation of the moons of Jupiter showed that there seemed to be a slight flaw in the celestial clockwork. In August 1676, Cassini noticed that the orbital period of the moon Io appeared to be somewhat longer as the Earth was moving away from Jupiter, and concluded that light would propagate at finite speed. Rømer then examined the data particularly carefully and dared to predict that Io would become visible ten minutes 'too late' on 9 November 1676. When this came true, he became famous overnight.

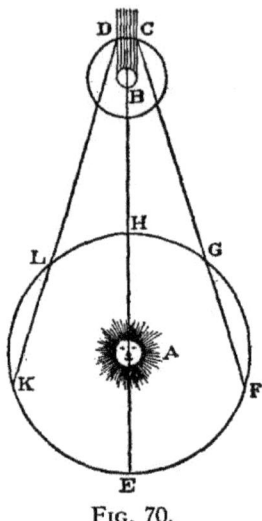

Fig. 70.

Schematic drawing by Rømer that illustrates the delay: The Sun is at A, and Jupiter at B. Io is orbiting Jupiter and therefore emerges from Jupiter's shadow at point D. However, during an orbit of Io, the Earth moves from point L to the more distant point K. The additional light travel time from L to K causes an apparent delay of the moment at which Io exits the shadow of Jupiter. The reverse effect is observed when the Earth moves from F to G.

The predicted ten minutes were actually fairly close to what the precise value for c would have suggested, but this was luck, too. For it was Christiaan Huygens who first calculated a concrete value for the speed of light, deducing 212,000 km/s from Rømer's data, which differs considerably from the true value 299792458 m/s. Surprisingly, though being an epochal discovery, finite speed of light was not accepted immediately. It took until 1725, when the so-called aberration of light provided additional evidence for a finite c: it was discovered that a telescope

8 Finite Speed of Light: The Subtle Anomaly

moving perpendicular to the direction of a star must be tilted slightly in order to have the propagating light aligned with the telescope's axis.[40] This is a nice example, however, that elucidates the interaction of astronomy with the rest of physics.

NOTHING HAPPENS FOR A LONG TIME

The delay in recognition was probably the fault of René Descartes, at the time the greatest scientific authority in France. He held fast to the dogma of an infinite speed of propagation, as did – perhaps against his own better judgment – Cassini. Isaac Newton, on the other hand, accepted Rømer's interpretation. Descartes was right about one thing, however: a finite speed of light did not fit into the system of philosophical simplicity that the natural scientists of the time were taking for granted. No one had ever predicted, let alone calculated, a finite speed of light. Rømer's discovery thus represented an anomaly, i.e. an observation that could not be explained within the framework of the established laws of nature. However, since it had little impact outside the realm of celestial mechanics, it was considered an astronomical curiosity rather than something fundamental.

Nothing changed that assessment for almost two centuries, before people conjectured – during the development of electrodynamics – that electromagnetic waves propagated at the speed of light. After Heinrich Hertz's famous experimental confirmation in 1888, it became obvious that the speed of light c played a paramount role in nature. Of course, Hertz's discovery, which at that time reduced the number of fundamental constants, was not perceived as an unpleasant anomaly, but was welcomed as unification. Back then, nobody had reason to question the role of c from an epistemological perspective. The speed of light, though still an unexplained quantity, had become a natural constituent of fundamental physics.

Part II: The End of Space and Time

SPEED OF LIGHT AND MASS

When Albert Einstein published his famous article *On the electrodynamics of moving bodies* in 1905, the speed of light suddenly became even more important. Besides the effect of time dilation that, according to the formula $\frac{t'}{t} = \sqrt{1 - \frac{v^2}{c^2}}$, predicted a slower time lapse of moving clocks, the impact on the dynamics of masses was of paramount importance. Einstein and others had realized that even with an unlimited supply of energy, a material body could never exceed the speed of light – a fact that put Newtonian mechanics into trouble. Accordingly, the mass of a body in motion increases by a factor

$$\frac{m}{m_0} = \frac{1}{\sqrt{1-\frac{v^2}{c^2}}},$$

as we have discussed in Chapter 3. What is truly remarkable here from a conceptual point of view, is that a finite speed of *light* not only exists, but it also represents a limiting speed for *matter*, which for some unknown reason cannot be accelerated beyond the value of c. These two fundamental phenomena of physics – light on the one hand and matter on the other – will accompany our search for the laws of nature as we proceed further.

In any case, the idea that the flow of time depends on the motion of the observer, and is therefore relative, must certainly be called one of the boldest ideas of the 20th century, if not of physics in general. There is no need to reemphasize the impact on the modern world view that the theory of relativity entailed. There is no doubt that special relativity provided insights that are indispensable for a deeper understanding of nature. Accordingly, relativity was celebrated as further development or even completion of Newtonian mechanics.

However, it has become quite forgotten that Einstein, just like Newton, did not explain the existence of the speed of light as

8 Finite Speed of Light: The Subtle Anomaly

such. From a historical perspective, this might seem to be asking too much, since Einstein, an unknown patent officer, had already had a hard time getting people to take his revolutionary theory seriously. Above all, the four-dimensional form, propagated by Minkowski from 1908 onwards, helped to divert attention from the yet missing reason for c's existence.

Initially, Einstein showed a certain ironic distrust of the four-dimensional version ("Since the mathematicians have invaded the theory of relativity, I do not understand it myself anymore."), but, unfortunately, no active resistance. Maybe because he was flattered by the success of the theory, it had escaped his intuition that this was the harbinger of a catastrophic derailment of theoretical physics: ignoring the fact that the speed of light c was still an unexplained phenomenon.

CONSTANT OR NOT?

An attentive reader of Chapter 4 may ask why the speed of light c is called a constant here. Einstein's thoughts of 1911, in combination with Robert Dicke's theory of 1957, however, showed that assuming the speed of light to be variable led to a convenient and intuitive formulation of the theory of general relativity. Thus, can a variable quantity be seen as a constant of nature? In a technical sense, yes. The fact that the speed of light depends on position and time is, from a methodological point of view, not a complication. Rather, it allows the gravitational constant (whose apparent value also varies) to be computed from the data of the universe. Therefore, one should rather speak of determinable quantities communicated by nature instead of 'constants' of nature. Nonetheless, they ought to be explained. Thus, instead of being a misconception, variable speed of light offers new opportunities to better understand the nature of c. For example, it seems promising to explore the true meaning of the expression $1/c^2$. On the one hand, it occurs in the gravitational potential of the universe; while on the other hand, it is proportional to cosmological time (cf. Chapter 4).

From the perspective of natural philosophy, it is essential that nature has set such a speed restriction in the first place, regardless of whether it can vary locally. Such an upper limit for the velocity of moving bodies is a property of nature whose existence cannot be justified a priori. Why are there no laws of nature that can do without it? This question has never been properly discussed. Psychologically, this might be understandable, because special relativity has provided a wealth of stunning explanations that led to deep insights. For example, the formula of kinetic energy, known since 1726, may be rederived from the relativistic mass increase.

> Consider the relativistic expression for the total energy $E = mc^2 = \frac{m_0 c^2}{\sqrt{1-\frac{v^2}{c^2}}}$. By using the well-known approximation $\frac{1}{\sqrt{1-x}} \approx 1 + \frac{1}{2}x \dots$ the above term results in $E = m_0 c^2 + \frac{1}{2}mv^2 + \dots$ i.e. the rest energy and the expression for the kinetic energy, $\frac{1}{2}mv^2$.

Since relativity was apparently a great success, also from a methodological point of view, why should one agonize over the origin of c? In fact, c is the 'oldest' constant of nature of all, even since 'Newton's' constant G was first measured in 1798. It is probably because so much time passed between the discovery of c and the recognition of its wide-ranging significance that the epistemological shortcoming of the constant c is easily overlooked. No one predicted its existence, and especially not the role it plays in Einstein's relativistic dynamics. Therefore, velocity-dependent masses contradict Newton's theory and actually falsify it. From Newton's perspective, there is no reason whatsoever why masses should not be accelerated beyond the speed of light.

The very existence of the speed of light is a blatant contradiction of classical mechanics. While this is generally accepted, the incompatibility is wrapped in cotton wool by stating that special

8 Finite Speed of Light: The Subtle Anomaly

relativity contains Newtonian mechanics as a limiting case. Indeed, the latter remains approximately valid in the low velocity regime. The mathematical consistency and elegance of special relativity glosses over the fact that the modification of Newton's laws had a price: the introduction of a new constant of nature c, or – to give it a less respectful name – an arbitrary free parameter.

THE MOST ELEGANT ANOMALY EVER

A single constant describing such a variety of physical phenomena is admittedly something one can hardly criticize severely. After all, there is a relative parsimony in the laws of nature that make use of c.

According to Thomas Kuhn, however, introducing an arbitrary parameter results in a complication that usually rescues an established theory from being abandoned completely – and in this case it was Newton's mechanics being saved. Following strict scientific logic, this is an unhealthy remedy, since postulating new parameters often merely masks the deficiencies of a model. The overwhelming complications of medieval astronomy still provide the best example.

It may sound somehow irreverent to draw parallels between Einstein's theory of relativity and the epicycles of the geocentric model. By comparison, relativity is mathematically more convincing and more effective in terms of 'quality' of a law of nature. However, if one takes a sober look at the facts and counts the constants of nature, it cannot be denied that the difference is ultimately gradual: the Ptolemaic picture dealt with a plethora of 'constants of nature'; Einstein's theory uses only one. But perhaps one too many.

It should be emphasized here that these arguments have nothing to do with the ubiquitous criticisms of special relativity one encounters when browsing random websites,[1] all of which I consider – let us be clear on this – to be unjustified. Rather, it is only that methodological aspect of an arbitrary constant (which has slowly revealed itself over centuries) that I find unsatisfactory.

Perhaps it seems too daring for many to see the theory of relativity as a mere workaround invented for the sake of not having to discard Newtonian mechanics; an attempt to save at least its important concepts such as velocity, force etc. Yet, if one wants to pursue natural philosophy without any belief in gods, the speed of light c must be considered a free parameter that has to be eliminated, possibly by a theory that provides deeper understanding. Though the speed of light c assumed its final role only after many generations of scientists, this should not hide the fact that c is an anomaly that falsifies the underlying theory. Although Newton's classical mechanics are the suspect here, there is nothing derogatory in recognizing an elementary inadequacy in this great intellectual achievement of mankind.

BACK TO RESET FOUR HUNDRED YEARS AGO

Newton's theory, on the other hand, is based on crisp logic and at the same time so frugal in its assumptions that only a few concepts remain in which we may look for defects. The only notions that Newton did not derive were space and time, whose existence he took for granted.

> *Absolute, true, and mathematical time, of itself, and from its own nature, flows equably without relation to anything external. – Isaac Newton*

---

[1] Among them are arguments invoking 'pure logic', often consisting of a few words that declare the measured phenomena of time dilation or mass increase to be impossible.

8 Finite Speed of Light: The Subtle Anomaly

For Newton, Euclidean three-dimensional space and a uniformly running one-dimensional time was so evidently real that he did not bother justifying these concepts, though they were the basis of his subsequent considerations. Yet it is exactly here where we have to stop and start over if we want to thoroughly understand the phenomenon of the speed of light. It is unacceptable that motion in space and time should obey an arbitrary restriction such as a maximum speed c. Rather, the apparent existence of a speed limit has to be deduced from a theory that, evidently, is still unknown. When exploring new mathematical grounds, we cannot help but leave the safe haven of our reality, the perception of space and time. They have been taken for granted since time immemorial, yet we should examine mathematical objects built on first principles that could possibly explain why our senses indulge in the illusion of space and time.

Part II: The End of Space and Time

9 Shrewd Atoms: Another Problem for Newton

Even before the role of the speed of light in physics was fully understood, another mysterious constant of nature was to turn the physics of the early 20th century upside down. As already mentioned, Max Planck introduced the quantum of action h into his radiation law, although he explicitly did not want to assign a physical meaning to it. The outstanding importance of h was revealed by Einstein, who postulated that light released energy in portions ('quanta') amounting to $E=hf$. Ultimately, Niels Bohr noticed that the angular momentum of an electron orbiting the atomic nucleus is always a multiple of $\hbar = \frac{h}{2\pi}$. The ensuing development of quantum mechanics would have been unthinkable without h, and consequently the quantum of action is sometimes considered the most fundamental of all constants of nature. Incidentally, Einstein was also involved in an important observation that was related to h.

To verify Bohr's bold assumption about the role of $\hbar$ as angular momentum, in 1915 Einstein and the Dutch physicist de Haas devised an experiment that should harbor a surprise when being performed. Even in a metal, electrons carry angular momentum, but due to their different orientations, it is usually impossible to measure. However, when the device is placed in an external magnetic field, electrons will align their angular momentum in its direction, as if the particles were being oriented by an invisible comb. Under these conditions, reversing the polarity of the magnetic field causes all electrons in the metal to invert the direction of their angular momentum. This collective behavior of the lightweight electrons can be measured as a small change in angular momentum of the entire piece of metal, which was attached to a sensitive torsion fiber.

BEING SEDUCED BY EXPECTATIONS

Einstein and de Haas had anticipated a certain result, but at the same time, performing measurements with the delicate apparatus produced error-prone results. Einstein was therefore satisfied with an outcome that agreed with the theory to within two percent, and rejected another measurement in which the magnetic effect was almost one half greater. In truth, however, as more thoroughly conducted experiments soon demonstrated, the magnetic moment was twice as large as initially expected! It would be desirable for this psychologically instructive error to be taken to heart by a larger fraction of today's scientists, who are often too focused on the confirmation of their pet theory when analyzing data.

The fact that the magnetic field turned out to be twice as strong as expected made no sense at all if one had assumed the classical picture of a rotating charge distribution. Clearly, this was an anomaly in an epistemological sense and is basically a problem that has remained unsolved to this day. Later, in an attempt to explain this strange behavior, Uhlenbeck and Goudsmit introduced the concept of spin – at the time, simply imagined as an electron literally spinning around its own axis. The idea is however incorrect in yet another way.[1] We will return to these problems in Chapter 12.

---

[1] As early as 1925, the Dutch physicist and Einstein's mentor, Hendrik Antoon Lorentz, had shown that a particle of such small size could not generate the required large magnetic moment unless it rotated at superluminal speed. Modern physicists would call such arguments 'outdated classical ideas' and claim that 'quantum effects' must be taken into account. Nobody knows what this means concretely, however.

9 Shrewd Atoms: Another Problem for Newton

From the perspective of natural philosophy, the first question to ask here is why such a phenomenon as spin exists at all. In 1928, Paul Dirac attempted to develop a form of Schrödinger's equation that was compatible with Einstein's theory of relativity. He constructed mathematical objects that correctly reflected certain properties of spin. Yet it is incorrect to state that they shed much light on the origin of spin or even explained it.[I] It remains a complete mystery that all elementary particles display this property. Why are there no microscopic objects in physics with perfect spherical symmetry, a form that would actually reflect the idea of a particle? We don't know. But even if one wants to see a justification for the phenomenon of spin in Dirac's calculations, they certainly do not provide a good reason for the occurrence of the fundamental constant h.

QUANTUM PUZZLES REBORN

The quantum h continues to surprise scientists to this day. Particularly in the regime of ultracold temperatures in which superconductivity is observed, there were several experiments in which h suddenly appeared in the data. For example, the British physicist Bryan Josephson was awarded the Nobel Prize in 1972 for having discovered surprising currents between different metals. The strength of these currents in these so-called Josephson junctions is related to h.[II]

Another Nobel Prize was awarded in 1986, when German physicist Klaus von Klitzing discovered that the electrical resistance in a so-called Hall sensor (a device for measuring magnetic fields) was always a multiple of h/e^2, i.e. quantized. Today,

---

[I] See also chapter 12.
[II] When applying a DC voltage U, an alternating current of frequency $f = 2\,e\,U/h$ is observed.

this effect is used for the precision measurement of the two fundamental constants e and h. In 1996, another Nobel Prize was awarded for the so-called fractional quantum Hall effect.

The constant of nature h may therefore proudly look back on an extraordinary 'career'. For an entire century, it has led not only to exciting discoveries but also to groundbreaking technology: just think of lasers, digital cameras and computer technology. This is probably the reason why its methodological downside – namely the role of h as an anomaly – has gone practically unnoticed. In fact, the occurrence of h was characterized precisely by phenomena that could not be understood within the realm of existing knowledge.

THE THEOREM OF IGNORANCE

The quantum of action seemed to have a certain predilection for contradicting existing knowledge. This becomes clear if we consider one of the most important quantum theorems based on h, Werner Heisenberg's uncertainty principle.[1] From 1913 onwards, the debate on how to interpret Bohr's undeniably successful atomic model dominated the world of physics for almost two decades. The primary actors were Einstein, Heisenberg, Schrödinger, and Bohr, who often engaged in discussions for hours, which incidentally testifies to a completely different culture of debate than is common today. As the anecdote goes, Bohr and Einstein were once so engrossed in conversation that they forgot to get off the street car and travelled back and forth through Copenhagen several times. When Schrödinger visited Bohr in 1926, Bohr dragged him into discussions until Schrödinger took to his bed – either because of a cold or having been talked ill.

---

[1] David Lindley's book *Uncertainty* offers a very appealing overview.

9 Shrewd Atoms: Another Problem for Newton

Allegedly, however, this did not stop Bohr from sitting on the edge of the bed and arguing: "Schrödinger, you must see…"

Heisenberg was following an interesting philosophical strategy he had first outlined, according to his own account,[41] during a conversation that took place in 1925 in Einstein's apartment in Berlin. Heisenberg claimed that certain processes, such as electrons orbiting the atomic nucleus, may simply not be precisely observable as a matter of principle, recalling that Einstein himself had focused on observable quantities while justifying relativity.

In a similar fashion, Heisenberg tried to gain a general view of the new problems, and limited himself to describing observations. He summarized his findings in the so-called uncertainty principle named after him. According to that relation, certain pairs of quantities, such as position and momentum, but also energy and time, cannot be simultaneously measured with any accuracy. If, for example, one tries to determine position with an accuracy of Δx, the momentum cannot be measured more precisely than $\Delta p = h/\Delta x$ for reasons of principle. The uncertainty relation is convincingly verified by experiment. For example, very short-lived particles (small Δt) show a correspondingly broad distribution of their energy amounting to $\Delta E = h/\Delta t$. The uncertainty relation applies to all pairs of quantities whose product carries the unit of h, Nms. Therefore, it can almost be regarded as a summary of the riddles surrounding h.

The uncertainty principle relates in an interesting manner to a well-known theorem of the mathematician Emmy Noether about symmetries and conservation laws in physics. According to Noether, demanding the equations of motion to be independent of position leads to the conservation of linear momentum in mechanics. Energy, on the other hand, must be preserved if one seeks time-independent laws of nature.[1] This is not the case at all in cosmology, and consequently the concept of energy must be suitably generalized.

Heisenberg's formula involving the constant h is undoubtedly a remarkable discovery, but as a law of nature it is special in one respect: it quantifies ignorance rather than knowledge. Although the uncertainty relation is in perfect agreement with all experiments that show the apparently random scatter of measurements, the regularity lies precisely in the deviation from the calculable. In this sense, the role of h is perhaps best illustrated by the uncertainty principle, yet it constitutes an anomaly that, from an epistemological point of view, forces us to question anything that has hitherto been assumed to be certain.

CLASSICAL PHYSICS HAS A PROBLEM

In particular, the uncertainty relation, indeed the mere existence, of h heavily contradicts Newtonian mechanics. Considering its classical concepts there is no reason whatsoever why it should be impossible to measure the momentum[II] and position of a body simultaneously. The same applies to the more intuitive notions of energy and time, which, according to the classical view, could coexist as precisely measurable quantities.[III] In addition, there is no apparent reason for the quantization of angular momentum. Quantum mechanics thus blatantly contradicts classical physics.

---

[1] Cf. the slow decrease of frequencies $f \sim t^{-\frac{1}{4}}$. Already in the 19th century, the German polymath Hermann von Helmholtz had raised the question of why energy occurs in two such distinct forms, kinetic and potential energy.
[II] Linear momentum is the product of mass and velocity.
[III] There was a legendary discussion on the subject between Bohr and Einstein at the Solvay conference in 1930. One evening, Einstein claimed to have refuted the uncertainty principle with a thought experiment, which caused Bohr to have a sleepless night. The next day, however, Bohr proved Einstein wrong: he had overlooked an effect of his own theory of general relativity.

9 Shrewd Atoms: Another Problem for Newton

In formulating quantum theory, however, great efforts were made to preserve classical mechanics as a limiting case, to avoid inconsistencies. However, it might have been more honest to regard quantum phenomenology as evidence that disproves Newtonian physics.

Just as classical mechanics continues to be valid for velocities that are small compared to c, there is no contradiction of Newtonian mechanics as long as the products of position and momentum, respectively energy and time, are large compared to h. Just as physics clashes with c on the large astronomical scale, it stumbles over h on the atomic scale. Expressed more formally, classical mechanics applies only in the two limiting cases in which the speed of light c is infinitely large and Planck's constant h infinitely small. Unfortunately, this is not the case in the real world.

If we reconsider the physical units in Chapter 7, we find an intriguing aspect of Heisenberg's uncertainty relation. If the unit of mass, i.e. kg, is replaced by an inverse acceleration s^2/m, the unit of h becomes m·s, i.e. the product of length and time. The obvious problems with the notions of space and time would thus reappear in Heisenberg's principle as a space-time uncertainty. Whether this observation leads to further insights, or whether the uncertainty relation itself is part of the problem, remains to be seen.

KUHN: IMMUNE TO MATHEMATICAL EXCUSES

The usual narrative of the history of physics is that around 1900, the constant h appeared from nowhere and explained a wide variety of experiments in the following decades. Methodologically, however, it is more correct to say that h was introduced into the theoretical apparatus to describe otherwise contradictory phenomena. Admittedly, this was achieved in a relatively parsimonious manner with only one single new constant, the quantum of action. Nevertheless, if one takes stock after a century of discoveries, h remains an arbitrary free parameter introduced to save an established theory that otherwise had to be abandoned completely. The quantum of action, in the sense of Thomas Kuhn, is therefore clearly an anomaly that points to a problem of the underlying model – Newtonian physics.

Another irony of history deserves to be mentioned here. Together with Gottfried Wilhelm Leibnitz, Newton was the inventor of differential calculus that used infinitesimally small quantities for the first time in a formalized way. Based on this groundbreaking work, mathematicians have developed concepts such as continuity and differentiability, which are often used to prove physical theorems. However, if one actually deals with the inconsistencies on a microscopic scale, mathematical physicists are easily satisfied with statements such as "quantum corrections are necessary" – whatever that means. But such parlance is nothing other than an excuse that conceals the fact that on small time and length scales, our description of reality collapses. Even in this respect, Newton's construct, on which the mathematical description of nature has been based for centuries, collides with reality.

CAN WE CALCULATE THE FUTURE OF THE WORLD?

Finally, there is another an epistemological aspect of quantum theory that, since its beginnings, has led to an exhausting number of debates both in oral and written form. We are talking about randomness in nature. Since Henri Becquerel's discovery of radioactive decay, it has been clear that the precise moment at which an unstable atomic nucleus will disintegrate cannot be predicted. Radioactive decay can only be described statistically, referring to a large number of atoms. This patently contradicted any mechanistic description of nature and put an end to the notion

9 Shrewd Atoms: Another Problem for Newton

determinism that had existed since Aristotle and was taken for granted by philosophers like Descartes.

The rapid development of atomic physics at the beginning of the 20th century provided many other examples where randomness ruled. For example, the point of impact of an electron passing through a narrow slit is determined by chance, as is the time an electron remains in an excited state of an atom before jumping to a lower orbit and emitting light. Erwin Schrödinger was so frustrated about the idea of these random discontinuities that he said to Bohr: "If we are going to stick with these damn quantum leaps, I regret that I ever had anything to do with quantum theory."[1]

Erwin Schrödinger (1887-1961)

---

[1] Allegedly, Bohr replied: "The rest of us are thankful that you did!"

Even better known is the critique of Einstein, who summed up his reservations towards random processes in nature with the iconic phrase [I] "God does not play dice!"

Experiments leave no doubt: some processes can only be described statistically. The epistemological question is whether there are compelling reasons for this somehow frustrating behavior of nature. One may argue that recourse to randomness is a defeat of the mind, which, in order to maintain self-esteem, might want to find an excuse by assuming that something is impossible to calculate in principle because one has not yet understood the very principle itself. Regarding randomness, I do not want to argue in such a radical way, although renouncing causality remains quite an unsatisfactory solution.

RANDOM CRAZINESS

However, even those willing to accept randomness as one of nature's characteristics must recognize that there are weighty arguments indicating that there is a problem with the fundamentals of the physical world, space, and time. Quantum theory has to assume that randomness manifests itself only at the moment of measurement. It is believed that[II] microscopic systems, such as electrons in an atom, are governed by a wave function that only allows the calculation of a probability for the particle to be in a certain state. In particular, the result of a measurement is influenced by the very same measurement. Some researchers have interpreted this to mean that there is no longer a reality independent of the observer. This has also led to numerous debates.

[I] Einstein later specified his reservations with the statement *probabilitatem esse deducendam, not delendam!* In other words: it may be that nature behaves in a random manner, but if it does, this must be deduced somehow.
[II] So-called 'collapse of the wave function', 'measurement problem' etc.

9 Shrewd Atoms: Another Problem for Newton

Erwin Schrödinger, a staunch defender of objective reality, conceived the now-legendary thought experiment involving a cat in a box, which, depending on a random event that may or may not crack a vial of poison, may or may not have been killed.

Schrödinger mockingly suggested that the cat must be in a quantum mechanical superposition of being simultaneously dead and alive before someone opens the box to see what is going on with the unfortunate animal. By reducing the literal interpretation of the wave function to absurdity, Schrödinger exposed the conceptual difficulties that become apparent when microscopic processes are transferred to the everyday world.

Only a fool would forgo the concept of a real world around us. – Erwin Schrödinger

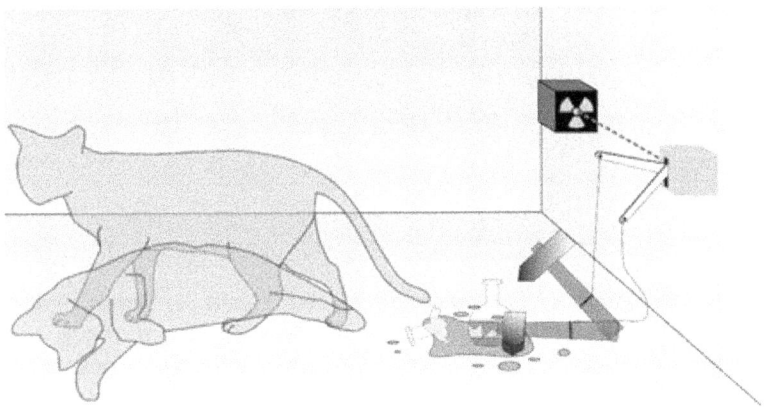

Sketch of Schrödinger's cat. A random radioactive decay leads to the opening of a vial filled with poison that may kill the cat.

ATTACK ON THE QUANTUM SPOOK

Einstein delivered an even harder blow to quantum mechanics by inventing yet another thought experiment. In 1935, already at Princeton, he had started a fruitful cooperation with the physicists Boris Podolsky and Nathan Rosen. According to the principles of quantum theory, so-called entangled systems exist. Consider the example of two electrons in the same orbital of a hydrogen atom, whose spins must be oriented opposite to each other, though each spin may choose its direction randomly. Therefore, once one spin is measured, the orientation of the other one is known *instantly*. The problem here is that such a system can be expanded in space, the two electrons being located at a distance. Hence, the information about one spin, telling the other one how to arrange, should actually be transmitted at superluminal speed as soon as the spin of one electron is measured. Einstein claimed this contradicted his theory of relativity.

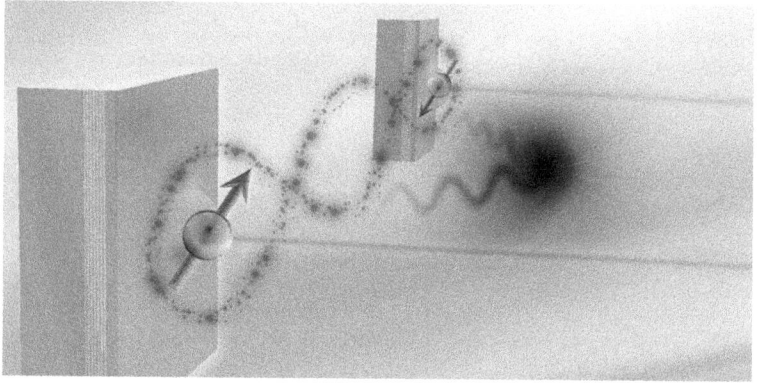

Schematic image of entanglement of two electrons. Although the locations of the possible measurements may be at a distance, the system is described by a single wave function. Yet the 'entangled' spins are always oppositely oriented.

NO ESCAPE EVEN FOR OPTIMISTS

Many decades later, the French physicist Alan Aspect ultimately succeed in proving that nature actually does behave so crazily: The electron spins felt their alignment immediately, even at a great distance.[1] Einstein was wrong.

---

[1] This is called *non-locality*.

9 Shrewd Atoms: Another Problem for Newton

This fascinating property of entanglement is believed to be a crucial element in the development of the quantum computer. Instead of conventional bits of binary value 0 or 1, a quantum computer uses so-called qubits that take on continuous values and adopt the states 0 or 1 only with a certain probability – just like the spin of an electron.

The theoretical discussion on how to understand these and related phenomena continues to this day, and it seems increasingly unlikely that quantum theory will ever be able to provide a consistent description of nature.[42] However, none of the common alternatives has delivered a logically satisfactory picture so far. Expecting that that the scientific community will ever agree on one of the known interpretations of quantum mechanics seems rather naive. Such prolonged stagnation suggests that the real cause of the difficulties has not yet been identified. If it turns out that space and time are indeed unsuitable concepts to describe reality, this would obviously result in many subtle contradictions, such as those we observe in the quantum world. Hence, the decades-long unsolved conceptual problems already provide a clear indication of the inadequacy of Newtonian mechanics and its basis, space and time.

Despite its deep-rooted difficulties, quantum theory still has many adherents who do not see any fatal contradictions in it. Thus, the theory is often accepted as a sufficiently consistent working hypothesis for the development of new theories. However, if one looks at the history of science from a methodological perspective, this hope dwindles. For even the most indulgent proponents of quantum mechanics must acknowledge that it is based on the unexplained constant h. A quantum theory without this fundamental constant is inconceivable. This fact alone, despite all the success in describing experiments, is an insurmountable obstacle to any progress. When dealing with science – i.e. something that does not rely on gods – one must get rid of the

Part II: The End of Space and Time

quantum of action. But physics has so far been unable to provide a deeper reason for the existence of h.

Obviously, the physical units of meter and second, inseparably linked to c and h, point to a problem with the notions of space and time. Any attempt to explain c and h, thus making them obsolete, must therefore question the concepts of space and time. If fundamental progress in the description of reality is to be made, the arbitrary 3+1-dimensional picture, despite being the basis of Newtonian physics, must be replaced.

The fact that the problems involving c and h are the most fundamental ones of theoretical physics does not mean, of course, that there are no other difficulties. Rather, Chapters 5 to 7 have shown that there is still a long way to go in eliminating the other constants of nature before turning to c and h. Whether those simplifications must be achieved first, or whether a deeper understanding of space and time might help to solve the previous problems, I cannot decide. In any case, the unsolved questions of physics must be visible in their entirety if we want to have a chance to understand reality.

Part III: The Mathematical Universe

"Time is said to have only one dimension, and space to have three dimensions. (...) The mathematical quaternion partakes of both these elements; in technical language it may be said to be 'time plus space', or 'space plus time': and in this sense it has, or at least involves a reference to, four dimensions. And how the One of Time, of Space the Three, might in the Chain of Symbols girdled be."

– *William Rowan Hamilton*

Part III: The Mathematical Universe

10 Possible Alternatives to Space and Time

The previous chapters have shown that both the speed of light c and Planck's quantum of action h are anomalies in an epistemological sense. They contradict the fundamentals of Newton's theory. But what could be wrong with it? Plainly speaking, classical mechanics is nothing but pure logic plus two axioms that Newton did not further justify: the existence of space and time. However, the occurrence of c and h proves that this is where the problem lies.

I have shown in earlier chapters that all other constants of nature can be eliminated as a matter of principle, but unlike those considered before, h and c are not computable. There is no conceivable equation, let alone a theoretical framework, from which one could derive the numerical values c=299792458 m/s or h=6.62607015·10^{-34} kg m²/s. Why is that? Simply because nature – apart from the properties of the universe and those of the elementary particles considered earlier – does not provide any further quantities from which the above values for c and h could be computed without falling into circular reasoning.

The numerical values themselves result from the man-made definitions of the scales of meter and second.[1] Does the arbitrariness of the above numerical values mean that one need not worry about c and h, as is often claimed? No way! Their mere existence – not the numerical values – poses a problem. Consequently, we cannot hope for an explanation by hitherto undiscovered coincidences, but rather must seek a *qualitative* justification for the fact

---

[1] In Chapter 7 it was explained that the unit kg should be expressed by an inverse acceleration, i.e. by m and s. Hence, the unit m results from $\sqrt{hc}$ and the unit s from $\sqrt{h/c}$.

Part III: The Mathematical Universe

that these two phenomena, c and h, occur in nature. Since the unjustified units of meters and seconds are measures of space and time, it is obvious that they are the concepts that demand scrutiny.

MATHEMATICS ALONE MUST SUFFICE

There is overwhelming historical evidence that scientific revolutions have led to simplification, and simplicity has proved to be a fairly good guiding principle for our philosophical considerations so far. But how can simplicity be achieved when describing nature? Ultimately, there is only one way: building a model of reality devoid of any 'physical' constants, i.e. justifying them by pure mathematics.[I] If we adhere to a rational philosophy of nature, fundamental constants must be seen as a consequence of our hitherto limited understanding. There is no epistemological reason that forbids describing the world around us in mathematical terms only, without resorting to arbitrary postulates such as 'constants of nature'.

Of course, this argument goes beyond Galileo Galilei's credo "The great book of nature is written in mathematical language", and it may surprise some readers of my earlier books that I am advocating here a theory based on pure mathematics for the description of nature.[II] For any real progress, however, c and h must be *explained*, i.e. manifest themselves as mathematical properties. The entire phenomenology of c and h must turn out to be a necessary characteristic of mathematical objects. And, of course,

---

[I] Though his line of reasoning is quite different, I agree with the central thesis of Max Tegmark's book *The Mathematical Universe* that external physical reality must consist of mathematical structures.

[II] I have criticized the pointless mathematization of physics by so-called 'string' theories, 'supersymmetric' theories or 'loop' quantum gravity, all the more because they are practically unfalsifiable. These theories not only do not care about simplicity, but have not even realized that the existing constants of nature call for an explanation.

the properties of these objects should also shed light on why we perceive them approximately as the 3+1-dimensional world commonly called space and time.

ARBITRARY DIMENSIONS

As the basis of his mechanics, Newton postulated a one-dimensional, uniformly running time and a three-dimensional, straight Euclidean space. This is called $\mathbb{R}^3$ because it uses the real numbers $\mathbb{R}$; moreover, distances between any two points can be computed. If one adds the time t, our usual model of reality is then called $(\mathbb{R}^3, \Lambda)$ with elements (x, y, z, t).

Obviously, this 3+1-dimensional construction of space and time is somehow arbitrary. Why the splitting up? Why a total of four dimensions? Why does the 'fourth' dimension, time, manifest itself so differently from the other dimensions? These questions cannot be answered within the conventional framework of physics. It is clear that one must also question the number of these dimensions that supposedly represent reality[1]. Recalling our guideline of simplicity, it would be interesting to consider one dimension less. The only thing we can be fairly sure about is that we are surrounded by a reality that is at least three-dimensional.

A reasonable strategy to find mathematical structures that could possibly describe reality would be as follows: start with the simplest objects, check whether they can account for the observations, and if they fail to do so, look at the next more complicated object that might have more dimensions. To summarize, the difficult task is to find a mathematical object that is simple

---

[1] Of course, we are not talking about 'string' fantasies which postulate many dimensions and then claim that most of them must be invisible. Richard Feynman once mocked this attitude: "String theory does not produce predictions, but excuses".

Part III: The Mathematical Universe

enough to be convincing from the perspective of natural philosophy, but at the same time contains sufficiently rich structures capable of reflecting the wide variety of natural phenomena.

THE FIRST STEP OF THE REVOLUTION

I shall disclose here that this strategy will lead to the so-called three-dimensional unit sphere, a manifold[1] with intriguing properties that are often unknown, even among mathematicians. It is also clear, however, that searching for a satisfactory picture of reality implies, to a certain degree, being speculative. Changing the world view of physics in a fundamental manner poses immense difficulties. Therefore, it is sometimes inevitable at this early stage to venture a guess in order to not miss a possible visionary idea. Any promising path should then be thoroughly examined.

The following considerations about the three-dimensional unit sphere are therefore aimed to be a first outline of a visionary idea that, if worked out, may one day replace the paradigm of space and time. The second step, setting up a mathematical model, is almost entirely absent, though in the next chapter we will try to get a glimpse of its foundations. However, there are so many properties of the three-dimensional unit sphere suggesting a connection to physical reality that it is difficult to believe in coincidence. In this respect, it is certainly justified to follow this path. On the other hand, there seem to be few alternatives if one wants to understand reality in a rational way. Naturally, the brainstorming process I propose here often appeals to imagination and intuition. Therefore, mathematicians who may consider some ideas half-baked or vague are explicitly asked for indulgence.

---

[1] A generic term for mathematical spaces such as such as lines, surfaces, 3-spaces and higher-dimensional objects.

Sloppy definitions sometimes help not to deter the popular readership, but of course the ultimate goal is to find more rigorous arguments that one day should replace these early thoughts.

It is inevitable that I should first introduce a few well-known mathematical objects. The reader who is familiar with them may of course skip those sections; however it is important to me that non-experts should have an idea of these mathematical structures and appreciate their elegance – particularly if they can be visualized effectively.

IN SEARCH OF SIMPLICITY

Which mathematical objects have been used to describe reality? What exactly do physicists call 'matter' and what a 'field'? Of course, a mathematical theory of reality cannot reflect the multitude of 'fields' currently used in physics. What is needed is a mathematical structure that is simple in every respect.

Another look at history would seem to be helpful here. An interesting attempt to establish mathematical simplicity was the aether theories that dominated theoretical physics for the entire 19th century. The (admittedly ambitious) goal was to understand electromagnetic phenomena by reducing them to the graspable laws of mechanics. Space was imagined to be an elastic continuum, like a block of (incompressible) rubber in which sound waves could propagate. Particles would be described as irregularities in the material.[1] Mathematically, the state of the aether was simply described by a displacement vector at each point

---

[1] In fact, a consistent picture of particles was integrated into the aether model much later, after the development of the theory of so-called dislocations in crystals in the 1950s. There are interesting parallels between dislocation theory and the unified theory by Einstein and Cartan in 1930. See my papers arxiv.org/abs/gr-qc/9612061 and arxiv.org/abs/gr-qc/0011064.

Part III: The Mathematical Universe

of $\mathbb{R}^3$ (the straight, 'Euclidean', three-dimensional space), indicating the shift of the material from the undisturbed to the deformed state. Different displacement vectors in adjacent points lead to distortions or discontinuities that can cause waves or describe dislocations (particles). Mathematically speaking, what is needed here is merely a so-called vector field: at any point (x, y, z) in three-dimensional space, you can imagine a small arrow whose direction is also given by three coordinates (v_x, v_y, v_z). This is called a map $\mathbb{R}^3 \rightarrow \mathbb{R}^3$. If one imagines a field that changes over time, it is a (time-dependent) map $(\mathbb{R}^3, \Lambda) \rightarrow \mathbb{R}^3$. This is a very useful notion, since it can describe various phenomena such as a velocity field in a fluid or electric and magnetic fields.[1]

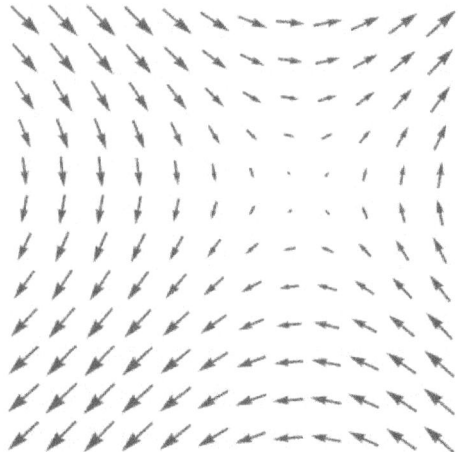

Sketch of a two-dimensional vector field. Correspondingly, one imagines a three-dimensional vector field by an arrow with direction and length at every point in 3-space.

The aether theories of the 19th century aspired to simplicity inasmuch as they sought to describe all natural phenomena by the displacement vector alone. Historically, it is interesting that many 19th century physicists, including the eminent Lord Kelvin,

---

[1] Of course, such a map $(\mathbb{R}^3, \Lambda)$ assumes the common 3+1 dimensions.

10 Possible Alternatives to Space and Time

dismissed the newly introduced electric and magnetic fields as arbitrary postulates. From the point of view of natural philosophy, they were probably right, because the electrodynamic quantities have set in stone one of the first schisms of physics, the division of the world into mechanical and electrical phenomena. One may well regard this as a crucial step towards complication.

With Maxwell's equations, however, the concept of the electromagnetic field became established, and from 1905 onwards, aether theories were gradually abandoned. It is true that without a proper picture of particles, the aether contradicted Michelson's 1887 experiment. However, this is not necessarily true in general.[1] Thus, the theory of an incompressible aether put forward by in 1839 by the Irish physicist James MacCullagh, which anticipated parts of Maxwell's theory, remains extremely interesting.[43] Nowadays, aether theories have been superseded by electrodynamics, whose key concepts, the electric and magnetic fields, are also described by vectors. Since light is an electromagnetic wave, it is fair to say that we also need vector fields to characterize light.

THE EVOLUTION OF NUMBER SYSTEMS

An abstract term that will be useful later is fiber bundle, a notion that includes vector fields as well as other quantities. One can imagine a fiber bundle similar to a hairbrush consisting of a wooden body (representing the base space or bundle, e.g.$(\mathbb{R}^3, \Lambda)$) and attached bristles (being called the fiber, here $\mathbb{R}^3$). Using this general term, our problem can best be formulated as follows: what is the simplest possible fiber bundle that is capable of providing a complete description of reality? Such an object should certainly resemble vector fields, but unfortunately this is

---

[1] Topological defects provide a consistent description of particles, and their motion even shows effects that are reminiscent of special relativity (C. F. Frank, Proceedings of the Physical Society A 62 (1949), p. 131).

Part III: The Mathematical Universe

not enough. Since the mathematical apparatus of quantum theory was developed by Schrödinger[I] in 1925, complex wave functions have been considered to be essential for the description of matter, in this case electrons. At the very least, this description by complex numbers is very successful. In order to appreciate their characteristics, we need to take a short detour.

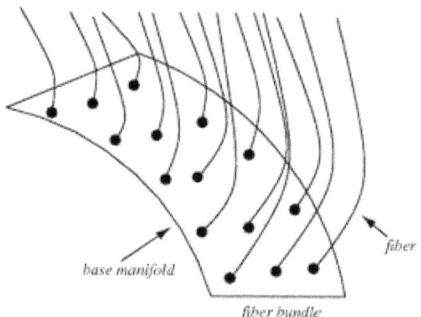

Schematic picture of a fiber bundle

The real numbers $\mathbb{R}$ form a one-dimensional continuum, often illustrated by an infinite continuous number line. Because one can both add and multiply real numbers in meaningful[II] way, $\mathbb{R}$ is called a field.

Extending this concept to two dimensions is straightforward: imagine a (two-dimensional) plane ($\mathbb{R}^2$) in which each point is determined by two real numbers (x, y), a pair that can also be imagined as an arrow with fixed length and direction.[III] Two 'arrows' can be added very easily,[IV] but it is not trivial to multiply

---

[I] Heisenberg's 1925 formulation, though devoid of the notion of a wave function, is equivalent to that of Schrödinger. Therefore, one may say that complex numbers occur in nature.

[II] This means there has to be a neutral element of addition, '0', different from the neutral element of multiplication, '1', etc. (so-called field axioms).

[III] Obtained from Pythagoras' theorem by setting $r=\sqrt{x^2 + y^2}$, and $\tan \alpha = y/x$ (for the direction).

[IV] By placing the tail of one vector on the head of the other; this corresponds to adding the respective coordinates.

10 Possible Alternatives to Space and Time

them in a way that a similar arrow, respectively a point in the plane, results. The system by means of which such a multiplication can nevertheless be meaningfully defined is called complex numbers (symbolized by $\mathbb{C}$ instead of $\mathbb{R}^2$) and was advanced by the brilliant Swiss mathematician Leonhard Euler (1707-1783). Initially, complex numbers were invented because people wanted to calculate the square root of a negative value. Therefore, the *imaginary* unit i was defined by exactly that property, $i^2 = -1$. This can be easily applied.

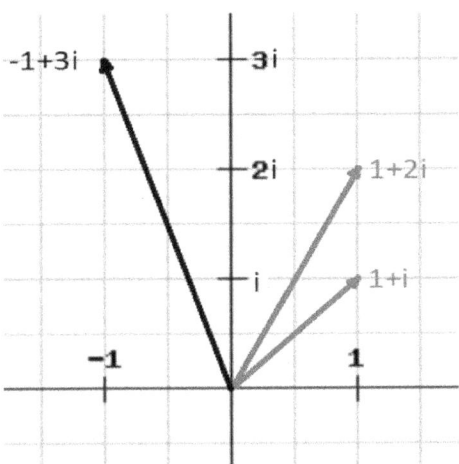

Multiplication of two numbers in the complex plane.

The complex numbers consist of a 'normal' real number and a multiple of i, the imaginary part. Again, they can be illustrated as arrows in the plane, y being the imaginary axis. When applying the convention $i^2=-1$, the multiplication can be carried out with common arithmetic (see illustration): The product of the numbers (1+i) and (1+2i), for example, results in (1·1+1·2i+i·1+2i²) =(1+3i-2)=(-1+3i). However, interpreting the result geometrically offers a surprise. The original factors, e.g. (1+2i), had 'length' (norm) $\sqrt{1^2 + 2^2} = \sqrt{5}$ and direction (from elementary geometry tan α = 2) α=63.4°, and these properties reappear quite beautifully in the resulting product! As one can easily verify, the

Part III: The Mathematical Universe

norm of the result amounts to $\sqrt{10}$, i.e. the product of the lengths of the original factors, but the resulting angle is simply the sum of the factor's angles! The two arrows simply added their counterclockwise rotations. This metamorphosis of a multiplication into an addition of angles is a particularly fascinating property of complex numbers. Demanding the multiplication in a perfectly 'straight' space like $\mathbb{R}^2$ to make sense inevitably leads to a kind of rotation. In this case, pure mathematics creates a complexity that we will encounter later in an even more pronounced way.

DO ATOMS NEED THE COMPLEX NUMBERS?

But why have complex numbers played such an important role in conventional quantum mechanics? Describing a particle by a wave function means that a complex number is assigned to each point in space where the particle can be at a certain time. Equivalently, we may speak of a map or fiber bundle $(\mathbb{R}^3, \Lambda) \rightarrow \mathbb{C}$. Why can simple real numbers not do the same job? In order to describe the presence of an electron in an atom correctly, one has to consider its wave nature. For such an oscillation, numbers are needed that can change their signs from positive to negative without becoming zero in the meantime.[I] As one can easily see, this is achieved in the complex number plane by moving in a circle around the origin $(0|0)$.[II]

This formalism, of which I have only given a brief overview here, is extremely successful when applied to the hydrogen atom and its electron shells, but fails to provide a simple picture if one takes into account other particles. This is because conventional

---

[I] The probability of detecting a particle is proportional to the square of the norm of the complex number. This number cannot be zero at any time, since then the particle would not exist.

[II] In fact, by multiplying with the number $e^{i\varphi}$ any rotation can be performed without changing the distance to the origin.

10 Possible Alternatives to Space and Time

quantum mechanics in this case requires a complex-valued wave function containing the coordinates of particles of all different kinds. The nonsensical multitude of particles in the standard model would inevitably reproduce itself if we follow that method. Instead, if we look for a solution that is satisfactory from the perspective of natural philosophy, only one single object may 'live' in a space-time point ($\mathbb{R}^3$, Λ) (if we want to retain this concept for the time being), even if that object might be a little more sophisticated than a complex number. For a single value, i.e. a map, ($\mathbb{R}^3$, Λ) → $\mathbb{C}$ is obviously not sufficient to represent the wealth of natural phenomena. Nevertheless, the description of matter must contain something similar to complex numbers.[1]

As early as the 1930s, the Dutch physicist and close friend of Einstein, Paul Ehrenfest, wondered[44] why the wave functions for matter (complex numbers) and light (vector fields) were mathematically so different. The importance of this profound question is still underestimated today. If one follows the mission to explain natural phenomena in a unified picture, light and matter must be contained in a single formalism. This means that there has to be a mathematical object that on the one hand, must be a little more complicated than vectors and complex numbers, but on the other hand must incorporate their properties. Both structures, vector fields, and complex functions can be regarded as fiber bundles, and we may therefore continue to use this term.

A FLASH OF GENIUS AFTER TEN YEARS

While we need a somewhat richer object than $\mathbb{R}^3$ (or $\mathbb{C}$) as fiber, at the same time we had also identified the bundle ($\mathbb{R}^3$, Λ),

---

[1] In his book *Space-Time Algebra,* the American physicist David Hestenes has proposed a fascinating interpretation of complex numbers. They can also result from real numbers in the context of differential geometry. Though a beautiful field of mathematics, the discussion here would take us too far.

Part III: The Mathematical Universe

i.e. space and time, as an arbitrary object that perhaps calls for a mathematically more elegant alternative.

William Rowan Hamilton, born 1805 in Dublin, was probably one of the most brilliant mathematicians of all time. A child prodigy who by the age of twelve had mastered just as many languages, he soon turned to mathematics and, of course, started to study complex numbers. If it was possible to define a multiplication in two dimensions in such an amazing way, was it also possible in three dimensions? Hamilton spent more than ten years of his life pondering over this question, and legend has it that for years, his son greeted him every morning with the question, "Well, Papa, can you multiply triplets?"

William Rowan Hamilton (1805-1865)

On 16 October 1843, while walking along the Royal Canal in Dublin, Hamilton finally came up with the answer. In three dimensions it was indeed impossible; but at that moment, he realized that the tricky multiplication of complex numbers could be

10 Possible Alternatives to Space and Time

transferred to a four-dimensional number system called quaternions that had three imaginary units i, j, k instead of just one i. Whether Hamilton could already have imagined the fascinating rotations that occur in this number system, we do not know. In any case, overjoyed at his idea, he carved the constituting equations into a stone of a nearby bridge:

$$i^2=k^2=j^2= i \cdot j \cdot k = -1.$$

This can still be read today on a commemorative plaque at *Brougham Bridge* in Dublin[1].

The author at Brougham Bridge in Dublin with a book on quaternions. Unfortunately, this historically important location has become somewhat unsightly today. Even the plaque had to be moved to escape the graffiti sprayers.

---

[1] Hamilton devoted the rest of his life to the study of quaternions, convinced that they would play an essential role in the description of nature.

Part III: The Mathematical Universe

While the name quaternion refers to four dimensions, they can be written in a way analogous to the complex numbers (a+bi). The straightforward notation for a quaternion is (a+bi+cj+dk), where a, b, c, and d are real numbers and i, j, and k the complex units mentioned before. The problem of multiplication Hamilton wanted to solve in three dimensions eventually required four dimensions. Nevertheless, there is a particularly interesting three-dimensional subset of the quaternions, which is defined by the relation $a^2+b^2+c^2+d^2=1$. In analogy to the complex numbers[1] of norm one, they are nowadays called *unit quaternions*, while Hamilton had dubbed them *versors*, indicating the occurring rotations.

A QUARREL: QUATERNIONS VS VECTORS

We can now understand for the first time why this object is called a three-dimensional unit sphere, abbreviated to S^3. According to the Pythagorean theorem, a circle is defined in two dimensions by the equation $x^2+y^2=1$; of course, the circle's line has only one dimension. Mathematicians who like to systemize, therefore call a circle a 'one-dimensional unit sphere' or S^1 for short. Analogously, a spherical shape akin to the Earth's surface is called S^2, which is defined in three dimensions by the equation $x^2+y^2+z^2=1$. Consequently, four dimensions are required to express the three-dimensional unit sphere by $a^2+b^2+c^2+d^2=1$. While four dimensions seem hard to imagine, in the next chapter I will present a trick that illustrates S^3 quite well. For the reader who likes to improve his mathematical skills, some algebraic rules with the quaternions have been listed in the gray-shaded boxes. It turns out that it is useful to split a quaternion (a, b, c, d) into a real part (a) and three imaginary components (b, c, d),[II] a view that simplifies

---

[1] The absolute value or norm of a complex number a+ bi is one if $a^2+b^2=1$ holds.
[II] Note that although a is arbitrarily chosen among the four components to act

10 Possible Alternatives to Space and Time

and facilitates the calculations, but is also noticeable for reasons of principle.

> The multiplication rule for quaternions reads as follows: $(a_1, b_1, c_1, d_1) \cdot (a_2, b_2, c_2, d_2) = (a_1a_2-b_1b_2-c_1c_2-d_1d_2, a_1b_2+b_1a_2+c_1d_2-d_1c_2, a_1c_2-b_1d_2+c_1a_2+d_1b_2, a_1d_2+b_1c_2-c_1b_2+d_1a_2)$, which seems a bit confusing. Interestingly, the product can also be written using the scalar and cross products known from vector analysis (see next box). If unit quaternion (a, b, c, d) is decomposed into a real part a and a vector $\vec{u} = (b, c, d)$. Then we get:
> $$(a_1, \vec{u_1}) \cdot (a_2 \vec{u_2}) = (a_2 a_2 - \vec{u_1} \cdot \vec{u_2}, a_1 \vec{u_2} + a_2 \vec{u_1} - \vec{u_1} \times \vec{u_2}).$$

Although one cannot directly identify this 3+1-dimensional structure (see box) of quaternions with space and time, it is interesting that this separation emerges for purely mathematical reasons. Incidentally, it was Hamilton who coined the term *vector* for the imaginary part (b, c, d) of a quaternion. Because the algebra of quaternions was not that easy, the vectors, although descended from quaternions, took on a life of their own.

A major proponent was the American mathematician Josiah Willard Gibbs who developed what is nowadays known as vector analysis (with the operators *div*, *grad* and *curl*). Moreover, vectors soon became important in Maxwell's electrodynamics as a practical tool for expressing electric and magnetic fields and began to dominate the literature. By contrast, quaternions still play a minor role today; the historical dispute about which is the more convenient tool has undoubtedly been settled in favor of vectors – which does not say much about their fundamental importance.

---

as the real part, the 3+1 structure inevitably arises.

Part III: The Mathematical Universe

The highly polemical style of the debate at the time is illustrated by the following statement:

> *"Quaternions came from Hamilton after his really good work had been done; and, though beautifully ingenious, have been an unmixed evil to those who have touched them in any way, including Maxwell."*
> *- Lord Kelvin, 1892*

In vector analysis, both the scalar product (also 'dot product') $\vec{a} \cdot \vec{b} = (a_1b_1 + a_2b_2 + a_3b_3)$ and the vector product (also 'cross product') $\vec{a} \times \vec{b} = (a_2b_3 - a_3b_2, a_3b_1 - a_1b_3, a_1b_2 - a_2b_1)$ are of significant importance and are widely used in most areas of modern physics. One may combine these products with the symbol of a spatial derivative, ∇, ("Nabla") and generate the differential operators *divergence* and *curl*. They consist of derivatives of the components (a_1, a_2, a_3) and indicate the source and vorticity (two quite intuitive terms) of a vector field, respectively. Maxwell's equations present the most prominent example of how fundamental laws can be formulated in an elegant vectorial form. Another operator (*gradient*) generates a vector from the three-directional derivative of a scalar field. Although vector analysis has some disadvantages compared to differential forms (another interesting calculus often used on manifolds), it has become the dominant notation in many applications. Just to illustrate the close relation between quaternion algebra and vector analysis, consider the quaternionic multiplication (see above) of a spatiotemporal derivative vector with electromagnetic potential:

$$\left(\frac{\partial}{\partial t}, \vec{\nabla}\right) \times (\phi, \vec{A}) = \frac{\partial \phi}{\partial t} - \vec{\nabla} \cdot \vec{A} \frac{\partial \vec{A}}{\partial t} + \vec{\nabla}\phi + \vec{\nabla} \times \vec{A}$$

The last two terms precisely match the known expressions for the electric and magnetic fields $\vec{E}$ and $\vec{B}$.

Perhaps it is of more than historical interest that Maxwell himself tried to formulate his equations by means of quaternions;[45] some authors have recently taken up these ideas.[46]

> *Take Hamilton's quaternions: the physicists threw away most of this very powerful mathematical system, and kept the part – the mathematically trivial part – that became vector analysis. But when the whole power of quaternions was needed, for quantum mechanics, Pauli re-invented the system in a new form. Now, you can look back and say that Pauli's spin matrices and operators were nothing but Hamilton's quaternions*[47] *– Richard Feynman*

FOUR OR THREE DIMENSIONS?

> *Somehow quaternions are a fundamental building block of the physical universe.*[48] *– William Hamilton*

If we come back to the philosophical question of what mathematical structure could potentially describe all physical phenomena, quaternions are a strikingly simple possibility. Since they contain both complex numbers and conventional vectors as a subset, quaternions, in principle, can represent all the number systems physicists have used in their description of the elementary phenomena light and matter. Here again, we can see the link to c and h, the unexplained constants of light and matter, whose imperfect understanding has even crept into mathematical formalisms.

If one restricts oneself among the quaternions to the unit quaternions, the mathematical property[1] Hamilton had been looking for is lost again; there is only multiplication, but no addition of

---

[1] Hamilton had sought to generalize complex numbers that constitute a *field*. To be precise, quaternions constitute a *skew field*, that means, a·b = - b·a holds.

Part III: The Mathematical Universe

unit quaternions. This, in turn, entails a number of subtle consequences that make the mathematical treatment anything but simple. Nevertheless, unit quaternions, equivalent[I] to the three-dimensional unit sphere S^3, are extraordinary objects whose peculiar properties are reminiscent of physical laws in many respects. We will come back to these coincidences later.

Precisely because the conventional four-dimensional 'space-time' is such a misleading concept, there is little hope of directly identifying it with the originally four-dimensional quaternions.[II] By contrast, the exceptional characteristics of quaternions that are reminiscent of physical phenomena, are already found in unit quaternions.

[I] To be precise, the two objects are homeomorphic. I will continue to use them as synonyms.

[II] Only the quantum mechanical wave function seems to be easier to realize by using the full four-dimensional quaternions. However, doubts remain as to how fundamental that formalism is.

11 The Three-dimensional Unit Sphere – Full of Surprises

In this chapter, I am going to present the fascinating properties of the three-dimensional unit sphere S^3 in more detail. This requires a certain focus on mathematical aspects, which are only later related to physical phenomena. Of course, the study of this simplest three-dimensional manifold is done with good cause. Indeed, for the reasons mentioned in the last chapter, it is obvious that S^3 may play an important role in the description of nature. We will see at the end of the chapter how the impression of a 3+1-dimensional space-time can also be created by a three-dimensional manifold like S^3 alone. Yet it is worthwhile concentrating on the mathematical properties first.

The originally four-dimensional quaternions were reduced to three dimensions by the equation $a^2+b^2+c^2+d^2=1$, considering only numbers with unit length. Similarly, S^1, a simple circle, was created from the complex numbers by imposing the condition $x^2+y^2=1$. Instead of the two arithmetic operations, addition and multiplication, only one is left now. It can be expressed either by adding angles (φ) on the unit circle S^1 or by multiplication[I] with the complex factor $e^{i\varphi}$. The latter point of view is preferable, as it is easier to generalize to elements of S^3 that can only be multiplied. Such a structure is called a group.[II]

One may now be concerned that S^3, being defined by an equation in four-dimensional space, will be practically impossible to visualize. Fortunately, this is not the case. It turns out that S^3 is

---

[I] This is evident from de Moivre's identity, $e^{i\varphi} = \cos \varphi + i \sin \varphi$.
[II] A group must fulfill three axioms: 1) a neutral element; 2) each element has to have an inverse a^{-1}, such that $a^{-1} a=1$; 3) the associative law $(ab)c = a(bc)$.

Part III: The Mathematical Universe

'almost' the same thing as a well-known instrument of mathematics, namely rotations in three-dimensional space, called SO(3). These rotations also form a group, which is rather intuitive: any two subsequent rotations can be expressed by a single one. Although many mathematicians and physicists are aware of the parallel between S^3 and SO(3), most of them, in my opinion, are not surprised enough by this amazing fact, which is anything but self-evident. It is astounding that S^3 so closely resembles an object of our conventional perception, and it will be even more interesting to see how subtly S^3 differs from the familiar rotations of everyday life. We shall take a closer look at those first.

LIKE A NORMAL ROTATION, ALMOST...

In three dimensions, any extended object can be rotated around three different axes.[I] On the other hand, a randomly oriented object in space can be returned to its original state by rotating it by a certain angle around a certain axis. This makes it obvious that the rotations in three-dimensional space form a three-dimensional number system.[II]

What will become important later is that the result of two consecutive rotations depends on the order in which they are performed. You may want to visualize this odd fact with an everyday object, say a book, by turning it successively by 90° around an east-west axis and north-south axis, while alternating the sense of rotation. It takes quite a while to get the book back to its original orientation. Give it a try! The same puzzling fact applies to elements of S^3. It is said that S^3 and SO(3) do not 'commute'. It

---

[I] In aviation this is known as *roll*, *pitch* and *yaw*.
[II] This is by no means a matter of course. Rotations in two-dimensional space can be described by a single angle, i.e. they are one-dimensional, while in a four-dimensional space, rotations around six different axes are possible, the axes corresponding to all possible pairs among the four coordinates.

11 The Three-dimensional Unit Sphere – Full of Surprises

now becomes clear that we have a tough nut to crack, because for 'normal' numbers 3 × 5 equals 5 × 3 (this is true also for complex numbers). However, this is what makes S^3 particularly interesting.

The common representation of rotations in three-dimensional space actually requires a number scheme of 3×3 entries, which is called a matrix. The box below contains a simple example of how to rotate the unit vector in the z-direction (0,0,1) around the x-axis by 90°, so that it points in the negative y-direction (0,-1,0).

> Multiplication of a matrix by a vector: The column vector is 'laid flat' and multiplied by the respective row of the matrix. Similarly, in matrix multiplication, a matrix is understood as three adjacent column vectors. Here is an example of a simple rotation by 90° (cos 90° =0 and sin 90°=1):
>
> $$\begin{pmatrix} 1 & 0 & 0 \\ 0 & 0 & -1 \\ 0 & 1 & 0 \end{pmatrix} \cdot \begin{pmatrix} 0 \\ 0 \\ 1 \end{pmatrix} = \begin{pmatrix} 1\cdot 0 + 0\cdot 0 + 0\cdot 1 \\ 0\cdot 0 + 0\cdot 0 - 1\cdot 1 \\ 0\cdot 0 + 1\cdot 0 + 0\cdot 1 \end{pmatrix} = \begin{pmatrix} 0 \\ -1 \\ 0 \end{pmatrix}.$$
>
> If the angle of rotation is not 90° but α, the matrix would read
>
> $$\begin{pmatrix} 1 & 0 & 0 \\ 0 & \cos\alpha & -\sin\alpha \\ 0 & \sin\alpha & \cos\alpha \end{pmatrix}.$$
>
> If, instead, three matrices are multiplied, which simulates successive rotations around three axes (each by the angle α, β, γ), the following expression results (so-called Euler matrices):
>
> $$\begin{pmatrix} \cos\alpha\cos\gamma - \sin\alpha\cos\beta\sin\gamma & -\cos\alpha\sin\gamma - \sin\alpha\cos\beta\cos\gamma & \sin\alpha\sin\beta \\ \sin\alpha\cos\gamma + \cos\alpha\cos\beta\sin\gamma & -\sin\alpha\sin\gamma + \cos\alpha\cos\beta\cos\gamma & -\cos\alpha\sin\beta \\ \sin\beta\sin\gamma & \sin\beta\cos\gamma & \cos\beta \end{pmatrix}$$

Any rotation, on the other hand, can be imagined as successive execution of three rotations around predetermined axes. If one performs the three corresponding matrix multiplications (the outcome also depends on the order), a rather complicated rotation matrix results, which was calculated for the first time by Leonhard Euler. From a computational point of view, however, nine numbers need to be stored if one wants to save a rotation matrix.

Part III: The Mathematical Universe

SIMPLIFIED ROTATIONS

Rather than by means of three Euler angles, an arbitrary rotation in three-dimensional space is easier to imagine if one thinks of just one angle φ and a properly chosen axis of rotation. The axis is defined by a unit vector with components (e_1, e_2, e_3), alternatively as a spatial direction with azimuthal and polar angles δ and θ, as practiced in astronomy. However, the amount of the rotation φ is not defined uniquely – rotations of 180° and -180°, for example, would lead to the same result.

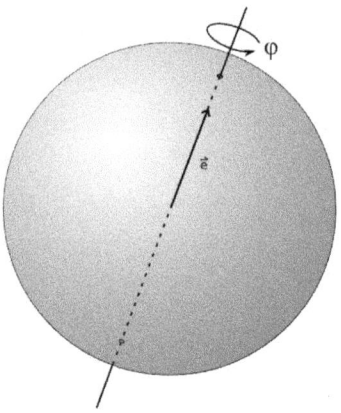

Representation of a rotation by an angle φ around a rotation axis defined by a unit vector (e_1, e_2, e_3).

Surprisingly, it turns out that the same information can also be encoded in the four components (a, b, c, d) of a unit quaternion, i.e. an element of S^3. In this case, the axis and angle of rotation can be read out much more easily than from the nine entries of the cumbersome Euler matrix. This offers enormous advantages for a number of technical applications (flight navigation and space travel, for example) or even for developers of computer games who have to deal with arbitrary three-dimensional perspectives; although this is, of course, not our primary interest here.

11 The Three-dimensional Unit Sphere – Full of Surprises

> A rotation in three-dimensional space around an axis with an angle φ can be represented by a suitable matrix multiplication. Such a multiplication transforms a unit vector $\mathbf{e} = (e_1, e_2, e_3)$ into another unit vector $\mathbf{f} = (f_1, f_2, f_3)$, φ being the angle between $\mathbf{e}$ and $\mathbf{f}$. However, this rotation can be executed more easily by a double multiplication from the left and the right with the unit quaternion $\mathbf{q} = (a, b, c, d)$, i.e.: $\mathbf{q^{-1} e q = f}$, whereby $\mathbf{q^{-1}} = (q, -b, -c, -d)$ is called the complex conjugate of $\mathbf{q}$. The axis and angle of rotation appear in an obvious manner in the following identity:
>
> $(a, b, c, d) = (\cos \frac{\varphi}{2}, \sin \frac{\varphi}{2} e_1, \sin \frac{\varphi}{2} e_2, \sin \frac{\varphi}{2} e_3).$ [49]
>
> When performing the double multiplication from the left and the right by $\mathbf{q}$ and $\mathbf{q^{-1}}$, the half angles of rotation $\frac{\varphi}{2}$ add up to the whole angle φ. We will come back to this important point later.

Even mathematicians who routinely compute such transformations might not have developed the kind of imagination that allows them to understand quaternion multiplication intuitively – which is not that easy in four dimensions, by the way. The visualizations created by computer expert Ben Eater are extremely valuable in this regard. It is not an overstatement to say that these simulations make you 'see' how spatial rotations actually occur by multiplying elements of S^3.

FOUR DIMENSIONS VISIBLE IN THREE

Let take a break for a moment from mathematical intricacies and direct our attention to an illustration tool that really helps to develop an intuitive understanding. To enjoy Ben Eater's videos, however, one needs to become familiar with a method that is also visualized effectively by Eater, but can also be explained here. It is called stereographic projection.

Part III: The Mathematical Universe

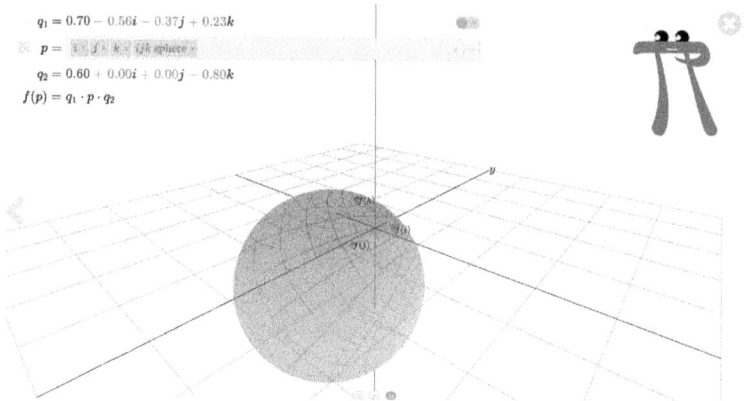

Visualization on Ben Eater's channel.[50] An unusual feature of these videos is that they are interactive, i.e. you can display examples of rotations by selecting your own parameters.

If one seeks a 'real' picture of the three-dimensional unit sphere, the problem to be overcome is that S^3 is defined by an equation in *four*-dimensional space, which is also called the 3-sphere's 'embedding', while our perception is unfortunately limited to three dimensions. However, cartographers have always had a related problem: how do you map the curved surface of the Earth, which is embedded in three-dimensional space, onto a plane, i.e. on the two dimensions of paper? Of the possible solutions, stereographic projection is one that is easy to understand, even if it is grossly distorting. Imagine the sphere being intersected by a plane that contains the equator. Now, if one pierces the sphere with a long straight needle through the North Pole and another given point of the Earth's surface, the needle or its continuation will pierce through the plane, too.

By this projection, the southern hemisphere, while being moderately distorted and reduced in size, is mapped onto the unit disk, maintaining a fairly realistic picture of the southern hemisphere countries. However, we now realize that the rest of the infinitely extended plane is reserved for the northern hemisphere, and it becomes clear that, for example, the projection of Greenland,

11 The Three-dimensional Unit Sphere – Full of Surprises

which is done at a very flat angle from the North Pole, would make it appear as a strangely warped area at a huge distance. Finally, the North Pole itself is not mapped anywhere, but identified with an 'infinitely distant point'. However, the stereographic projection gives you an approximate picture of the Earth's surface that no longer requires the third dimension, albeit at the cost of a huge distortion.

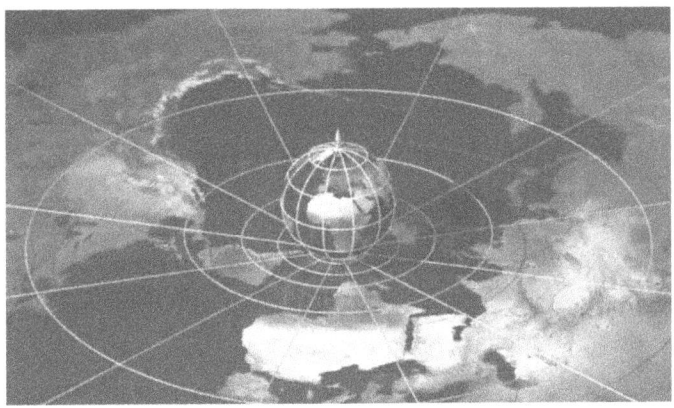

Two-dimensional version of stereographic projection, depicted in in a modified way. Here, the plane touches the South Pole (instead of cutting the Earth at the equator), which makes no big difference.

The principle behind the stereographic projection can be applied to any dimension, which includes illustrating the S^3. Since this is one dimension more than the S^2 considered before, we will need the entire three-dimensional space to visualize it, not just a two-dimensional plane. Yet it is helpful to remember that the curvature of S^3 has to be flattened out to fit in our Euclidean three-dimensional space $\mathbb{R}^3$. If we apply the above system analogously, the solid ball[51] around the origin (0|0|0) corresponds to the equatorial disk in the above two-dimensional projection. Hence, this ball will be, under the projection, a fairly accurate image of the 'southern hemisphere' (defined by a<0). Accordingly, the 'northern hemisphere' of the S^3 is being projected onto the rest of Euclidean space, the complement of the unit ball that surrounds the origin. As an aside, it is interesting that the mapping of the stereographic projection is conformal, i.e. preserves angles (a property we cannot prove here). This means that the circles in the S^3 are mapped onto circles in Euclidean space $\mathbb{R}^3$.

HALF THE TRUTH

A few further examples are helpful. The quaternion number (-1, 0, 0, 0) corresponds to the origin, and the three complex units i, j, and k simply represent the conventional unit vectors in the x, y and z directions. It is now easier to realize[1] why in the quaternion representation the rotation angles amount to only half the value of the 'true' angle in three-dimensional space. The quaternion representation consists of a pair of complex rotations, each complex rotation again consisting of two regular rotations. The complex rotations partly compensate each other, while the other parts add up to the full angle.

The fact that multiplications in the S^3 can be interpreted as the combination of two complex-valued rotations, [52] leads to another interesting representation of the S^3.

Complex-valued rotations are called elements of SU(2) and can be written as a complex-valued matrix of the type

$$\begin{pmatrix} a + bi & c + di \\ -c + di & a - bi \end{pmatrix}$$

which has the determinant 1. This sounds more frightening than it is, since, because of $i^2 = -1$, it means only that the equation $(a+bi)(a-bi)-(c+di)(-c+di) = a^2+b^2+c^2+d^2=1$, holds, the original definition of S^3. Hence, the 2×2 complex matrices of determinant one, SU(2), are equivalent to the three-dimensional unit sphere S^3 – the more intuitive notion I will continue to use.

---

[1] Once again, I emphatically recommend the animations by Ben Eater.

11 The Three-dimensional Unit Sphere – Full of Surprises

Let us get back to one of the basic properties of S^3, which can also be understood in an abstract way. Every real rotation in three-dimensional space corresponds to *two* points on S^3. For example, the numbers (a, b, c, d) and (-a, b, c, d) represent the same rotation, since they are each other's conjugate. Hence, in a fascinating way, S^3 describes all the usual rotations SO(3) twice. Mathematicians call this a 'double cover', a property whose physical interpretation will later catch the eye. Before that, however, it is worthwhile taking a closer look at why S^3 is not the same as SO(3) and at what the precise difference between the two manifolds is. I strongly recommend that you perform the following exercise yourself:[53]

THE PLATE TRICK

Hold a plate on the palm of your right hand in front of you and, for the sake of simplicity, consider only rotations around a vertical axis so that nothing can fall off the plate. Now rotate the plate by 360 degrees, i.e. in counterclockwise motion, pull it first to your hip and continue rotating it until, after wrenching your arm backwards and outwards, you bring it back into its original position. Now you will find yourself in an extremely uncomfortable posture with an inbound twisted arm that will soon cause your muscles to cramp. Can you imagine rotating the plate counterclockwise for another 360 degrees? It appears that this would injure your elbow, but amazingly, you can complete the task quite easily. Relieve your uncomfortable position by leaning your upper body back, lift the plate a little, and now continue the anticlockwise rotation above your head, after which you can readily complete the second full turn. You are holding the plate on your palm as you did when you started, but you have now experienced the painful difference between the rotations in S^3 and in SO(3)! The initial position and the uncomfortable position in the middle

Part III: The Mathematical Universe

of the process corresponded to one single element of SO(3), because the plate, having been turned by 360 degrees, was in the exact same orientation. However, that state was realized by two different elements of S^3: the comfortable and the uncomfortable position! You understand now that a full turn in S^3 corresponds to 720 instead of 360 degrees.

ENTANGLEMENTS AND RESOLUTIONS

If you keep this experiment in mind, you will better understand the most important tool mathematicians use to distinguish between manifolds: the so-called connectedness or, phrased more abstractly, the first homotopy group. For purposes of illustration, let us consider a closed path on S^2, the usual spherical surface (see picture).

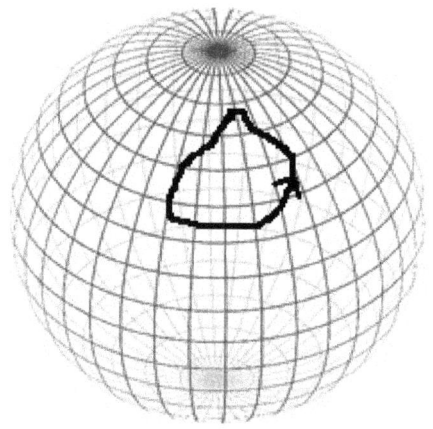

A two-dimensional spherical surface is called S^2. The picture shows a closed path on S^2 (with orientation) that can be shrunk to a point.

Any closed path on this surface can be contracted to a single point in a continuous manner, i.e. by a sequence of small changes. You may take this for granted, but the following figure shows a two-dimensional surface that is only slightly more complicated –

11 The Three-dimensional Unit Sphere – Full of Surprises

a torus, on which such a steady contraction does not work in general. There are even two distinct types of paths that cannot be contracted: those that go around the inner hole and those that surround the torus at its thinner circumference.

Each type can be characterized by how many times it circumnavigates the torus in a certain orientation.[1] The homotopy group is therefore said to consist in the 'direct product' of two integers: $\mathbb{Z} \times \mathbb{Z}$. On the surface of a sphere S^2, on the other hand, all paths are of the same kind, which is denoted as a 'trivial' homotopy group.

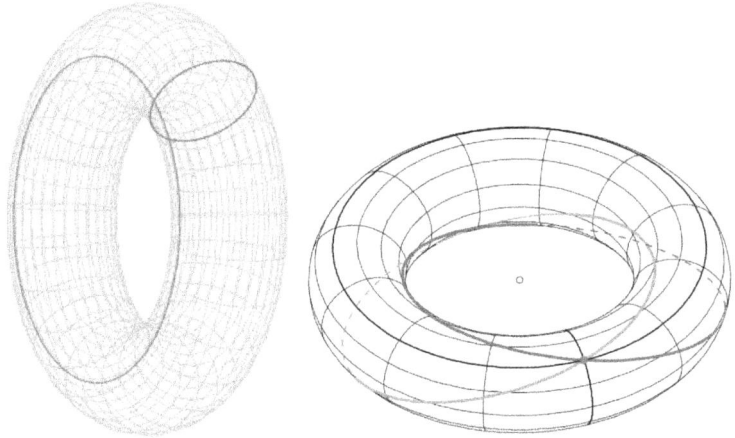

Torus with two essentially different, non-contractible paths (left). Combination of these paths (right).

What do we learn from this? Obviously, only simple manifolds have a trivial homotopy group, yet this group can become complicated quite quickly, even if the manifold is not that exotic.

---

[1] Thus, each path is characterized by two integers, for example (-3;5), which means: winding a path three times counterclockwise around the big hole and five times clockwise around the small circumference, and then connecting it again.

Part III: The Mathematical Universe

Now let us recall the plate experiment: As you can notice by contemplating the uncomfortable position, the group of rotations in the three-dimensional space SO(3) is simply not contractible, because your arm represents a closed path on SO(3). From shoulder to hand, each part of the arm was twisted by an increasing angle (in fact your elbow by about 180 degrees), but the end points, shoulder and palm, were in their original position. Since your arm is therefore the exact equivalent of a closed path, you do better not to try to 'contract' it. What would be a surgical emergency is, fortunately, simply impossible in math. Since you cannot bring it back into the comfortable position by constant deformation, that path is called 'non-contractible'.

The surprising fact, however, that the two consecutive rotations performed in the same (anticlockwise) direction did not cause a severe deformation of the arm, but resolved to an undisturbed state, also has a canny mathematical equivalent, namely 1+1=0, i.e. two identical (!) rotations neutralize each other. This equation is well known in algebra and belongs to the simplest number system that consists of 0 and 1 only (which, incidentally, is used by computers) and is known as $\mathbb{Z}_2$. It is therefore said that $\mathbb{Z}_2$ is the homotopy group of SO(3). This means that there is only the 'comfortable' and the 'uncomfortable' state, but no further possibilities of winding up your arm like on a torus. If, on the other hand, the rotations are described by S^3, the original state (1,0,0,0) and the 'uncomfortable' state (-1,0,0,0) are just two distinct numbers, and the path would not be 'closed' yet. If a path on S^3 actually starts and ends at the same point instead, it can also be continuously contracted. Hence, S^3 has a trivial homotopy group or, equivalently, is 'simply connected'.

TOOLS FOR THE PROBLEM OF THE CENTURY

You may wonder why I have gone into such detail in describing a mathematical subtlety such as the double cover. However,

11 The Three-dimensional Unit Sphere – Full of Surprises

it really seems that nature realizes this strange double representation of rotations. Moreover, one of the most important mathematical breakthroughs of recent decades was related to this double cover and the concept of homotopy. Henri Poincaré, the famous French mathematician, whose contributions to special relativity were of similar importance to Einstein's, first conjectured in 1904 that the three-dimensional unit sphere S^3 was the simplest three-dimensional manifold. To phrase it more precisely, there is no other three-dimensional manifold on which any path can be contracted to a point. The problem remained unsolved for a century, until the Russian mathematician Grigori Perelman published a proof in 2003, which was thoroughly examined by his most respected colleagues and accepted as correct. However, the eccentric genius rejected both a prize offer of one million US dollars and the even more prestigious Fields Medal.

Henri Poincaré (1854-1912)

Although this long-sought result was a greater revolution for mathematics than for physics, it is nevertheless important for our natural-philosophical considerations, because now we can be

Part III: The Mathematical Universe

sure that S^3 is indeed the simplest three-dimensional manifold. Once we consider it relevant for fundamental physics, this is satisfactory from an epistemological point of view. On the other hand, SO(3), the rotations in three-dimensional space, which is the common reality for us, is *not* the simplest three-dimensional object.

The Poincaré conjecture concerns topology, a branch of mathematics that deals with global properties of manifolds. Such topological characteristics do not change, even when the manifold undergoes arbitrary deformations. Rather, the topology of a manifold may only be altered by cutting and gluing. Perelman used in his proof the so-called *Ricci Flow*, an interesting concept which obeys equations that are similar to heat conduction in physics. Instead of temperature, however, curvature diffuses throughout the manifold; in other words, the manifold strives to equalize its varying curvature and to bring it to a constant mean value over time. It is remarkable that, even if there is no mention of physical time, a purely mathematical property of S^3 is reminiscent of the concept of time. In addition to the *Ricci Flow,* there are a number of other geometric flows that have to do with the evolution of curvature.

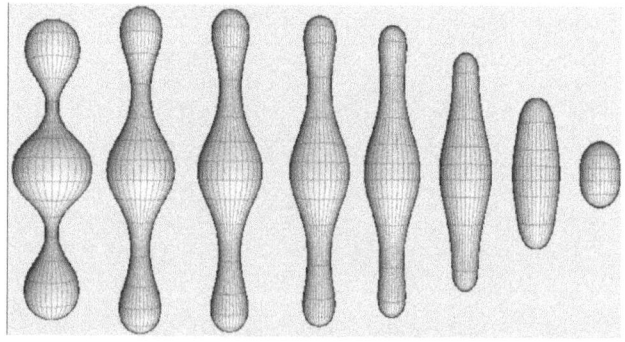

Illustration of the Ricci Flow: The curvature "flows" from the regions of large curvature to those with smaller curvature until a uniform curvature of the manifold (in this case the sphere) is reached.

11 The Three-dimensional Unit Sphere – Full of Surprises

Let us try to summarize what we have learned so far: The unit quaternions, or equivalently, S^3, contains both complex numbers and vectors, the concepts that are used in conventional physics to describe light and matter. S^3 is very similar to the rotations in three-dimensional space SO(3), and thus might generate the impression of living in a three-dimensional Euclidean space $\mathbb{R}^3$. Nevertheless, S^3 is even simpler than SO(3) in mathematical terms; indeed it is the simplest of all three-dimensional manifolds. Nevertheless, the properties S^3 show a surprising complexity.

WINDING PATHS ON S^3

Long before Perelman's proof, S^3 had already caused a stir among topologists. Let us visualize S^3 again as a rotation in three-dimensional space, while keeping in mind the small difference that a full turn corresponds to 720°, not 360°. However, we may as well map these 720° to a properly partitioned circle, i.e. the S^1. Each point on that S^1 would then correspond to a certain rotation angle φ, while the direction of the axis of rotation is determined by a point on S^2 (just as a star's position in astronomy is determined by two angles). We have therefore discovered a new, quite intuitive description of S^3, namely a S^2 to which an S^1 is glued at every point.[1]

Topologists love such decompositions that facilitate analysis, not least that of the homotopy group. However, in 1932 the German mathematician Heinz Hopf discovered that this decomposition into S^2 and S^1 does not work so smoothly. Naively, one might think that the S^1 circular loops could all be disentangled from each other if the surface of the sphere S^2 were cut open and laid

---

[1] A very simple parallel of this statement would be the following description for someone who has difficulties imagining a rectangle: "a rectangle is a width attached to each point of a length".

Part III: The Mathematical Universe

out on a plane.[1] In reality, however, all the circular loops were intertwined. Not a single pair could be separated without cutting one of them. This fascinating feature is called the Hopf Link.

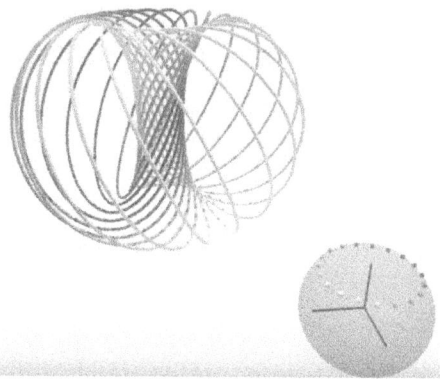

Illustration of the Hopf Link. At the top left, the entangled paths that correspond to circles in the stereographic projection; at the bottom right, the points on the S^2, each one a starting/end point of a loop.

> The Hopf map, or Hopf fibration, also has to do with the fact that every three-dimensional rotation can be represented by elements of S^3. Let us consider the product **q r q^{-1}**, where the amount of rotation is equally shared (remember the half angle $\varphi/2$) between the quaternion **q** and its complex conjugate[II] **q^{-1}**. **r** symbolizes a pure[III] quaternion, i.e. we may imagine it as a unit vector on S^2. Hence, a given rotation **q** transforms any vector **r** on S^2 into another vector on S^2, an obvious property of rotations SO(3). However, if we change perspective and look at a random element of S^3 that can be written as **q r q^{-1}**, we find that this map projects it back onto **r** that lies on S^2. This is exactly the Hopf map that, by reducing every loop to its starting/endpoint, shrinks S^3 to S^2.

---

[1] There is a very good computer animation of Niles Johnson (a mathematician from Ohio State University Newark) on YouTube: Hopf fibration – fibers and base.

[II] Only for *unit* quaternions the complex conjugate is equal to the inverse **q^{-1}**.

[III] This means of the form (0, b, c, d), i.e. without a real part.

11 The Three-dimensional Unit Sphere – Full of Surprises

If one studies the subject of Hopf fibrations in more depth, one will learn about analogous constructions in higher dimensions such as the S^7, i.e. the seven-dimensional unit sphere. Interestingly, it can be split into S^4 and S^3 by another Hopf map! This, of course, raises the question why we have limited our considerations to S^3 in the first place. On the one hand, S^7 (also known as unit octonions) is the logical continuation of the sequence $S^1 \rightarrow S^3 \rightarrow S^7$, since meaningful calculations[1] can still be carried out on these manifolds. On the other hand, it is not exactly an object that is philosophically appealing due to its great simplicity.[54] Therefore we will limit ourselves to the S^3 in the following.

NEW HORIZONS: HOW THREE DIMENSIONS CREATE THE ILLUSION OF A FOURTH

We have seen that S^3 can potentially replace both vector fields and the complex-valued functions. However, recalling the concept of a fiber bundle, this possible replacement related to the fiber, but it would still be a map $(\mathbb{R}^3, \Lambda) \rightarrow S^3$. However, the whole purpose of our considerations was also to question the Newtonian concept of space-time, i.e. the conventional bundle $(\mathbb{R}^3, \Lambda)$. If only the fiber is replaced by S^3, an arbitrary 3+1-dimensional space-time would still be unsatisfactory. Since we have adopted the perspective of natural philosophy, the working hypothesis of simplicity suggests here that not only the fiber but also the bundle should be replaced by S^3, i.e. reality is possibly described by a map such as $S^3 \rightarrow S^3$. But how is this going to work? An obvious objection is that a three-dimensional object is

---

[1] S^7 is a so-called divisional algebra, which is no longer associative [a(bc)≠(ab)c] and can therefore not be represented by matrices. In general, the higher the dimension, the more algebraic rules are broken, e.g. no ordering (complex numbers), no commutativity (quaternions), no associativity (octonions), etc. Occasionally, S^{15} and S^{31} are contemplated.

Part III: The Mathematical Universe

incapable of representing the four-dimensional phenomenology of space and time as a matter of principle. One dimension is lacking. However, this turns out to be too short-sighted an argument.

To understand this, we consider the more easily conceivable S^2 and a plane ($\mathbb{R}^2$), which in principle can be attached to any point of the curved surface of sphere. In the following, we call this plane the horizon space[1] of a given point. Just imagine what the manifold would look like if you flattened the curvature. The horizon space corresponds to the unobstructed view that is offered at certain points on the Earth's surface. There, we hardly notice the Earth's curvature and describe it approximately as a plane (a chart of an atlas!). We call this plane the horizon space of the observation point. Let us now imagine ourselves as flatlanders that cannot grasp three dimensions and walk around on the two-dimensional surface of a sphere. During this hike, we perceive a sequence of two-dimensional impressions from the respective horizon, which one could subjectively perceive as the passage of time.

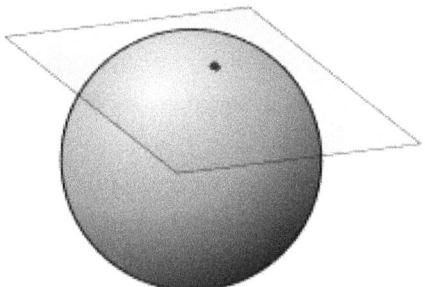

Example of a two-dimensional horizon space of a point on S^2.

---

[1] Not to be confused with the mathematically well-known concept of a tangent space, to which we return later. In differential geometry, what we call horizon space is often referred to as a chart of an atlas, a term that might mislead intuition however.

11 The Three-dimensional Unit Sphere – Full of Surprises

If this idea is applied in an analogous manner one dimension higher, it is imaginable be that we do not 'really' live in four dimensions, which means that time does not exist at all. Four-dimensional space-time would turn out to be a misconception. Rather, it is conceivable that a sequence of three-dimensional horizon spaces, for example on a curved manifold like S^3, appears to us as a movie that we only interpret as time[1]. But moving along a given path on S^3 does not prove that there is something like a fourth dimension; it just means that S^3 is curved and there is a path. These rather trivial facts may simply create the illusion of a fourth dimension. That being said, the idea of a sequence of horizon spaces is, unfortunately, still far away from explaining the phenomenology of space and time. As we shall see, the analogy cannot be complete for various reasons, but for the time being it may serve as a motivation to deal with this interesting property of the horizon space.

Just as you can imagine a plane glued to any point of the S^2, every point of the S^3 has a corresponding horizon space, which is of course three-dimensional. The horizon space is an excellent approximation for S^3 in the neighborhood of the point, while it deviates significantly at long distances – just as an attached plane fails to properly map the sphere at long distances. If distance measurements are taken in the neighborhood of a point on the S^2 or the S^3 (the latter distance corresponding to an angle of rotation), not only would similar values be obtained, but the angle would also be denoted in the same units as in the horizon space. In short: the horizon space looks very similar to a $\mathbb{R}^3$.

---

[1] Julian Barbour, in his book *The End of Time*, argues in a similar way: "time is change".

Part III: The Mathematical Universe

THE TANGENT SPACE – DEALING WITH SMALL CHANGES

The horizon space is to be distinguished from the so-called tangent space, which is often visualized in a similar fashion, but contains other units. Here is an example that illustrates the tangent space of the group SO(3), which is, unlike the group itself, identical to that of S^3.

Elements of SO(3) can rotate an object, say a ball, to any given orientation in three-dimensional space.[1] The situation is different if I only want to initiate the rotation by transferring angular momentum to the ball with a sudden twist. I can even add angular momentum around another axis simultaneously, because the momenta would add up like vectors (whereby the direction corresponds to the axis of rotation and the length to the amount of angular momentum), without the order being important. Thus, the elements of the tangent space can be imagined as angular momentum vectors or, more abstractly, as infinitesimal rotations. They are qualitatively very different from the distances in the horizon space, which have the meaning of a finite rotation angle.

Hence, the concept of tangent space is very useful when considering small changes and thus predetermined to express the process of derivation. For derivatives, i.e. infinitesimal rotations in S^3, different algebraic rules apply than for finite rotations. A long time ago, the Norwegian mathematician Sophus Lie developed a formalism to describe these subtleties. Groups like S^3, in which it is reasonable to consider infinitesimal changes, are called Lie groups or differentiable groups, in honor of Lie's achievements.

---

[1] When performing two consecutive rotations, however, the order becomes important, because SO(3), as S^3, is not commutative.

11 The Three-dimensional Unit Sphere – Full of Surprises

LIE GROUP AND LIE ALGEBRA

The infinitesimal rotations themselves, which add up like vectors, form the tangent space, which practically looks like a Euclidean space $\mathbb{R}^3$. This vector space of infinitesimal changes is called Lie *algebra*. This is significantly different from 'conventional' mathematics. A function of real numbers such as $f(x)=x^2$, for example, does not change its character as a function of real numbers when taking the derivative, namely $f'(x)=2x$.

This type of differential calculus is widely used and characterized by the fact that the derived objects are of the same quality as the original ones.[I] This is quite different in the case of S^3. If one considers S^3-valued functions, their derivatives are elements of the Lie algebra so(3) (lowercased). S^3 actually has the same algebra as SO(3), because the double cover (which distinguishes S^3 from SO(3)) is a global property that is lost when looking at infinitesimal elements. The derivative of S^3 is thus a qualitatively different mathematical object than S^3 itself. If we think about applications, this would mean that we have to be careful trusting differential equations – which make up a sizable part of theoretical physics. If some physical quantities should turn out to be S^3-valued, then the known equations would be an approximation at best. This will become interesting later.

Many well-known concepts of vector analysis, in particular differential and integral calculus, cannot be transferred easily to Lie groups.[II] Rather, there is quite a lot of unexplored territory in mathematics. Nevertheless, there are certain neat mathematical

---

[I] This is at least true in a formal way. If the derivatives are interpreted correctly, e.g. as slopes, their numerical values certainly cannot be added in the usual way. However, this is not entrenched in formalism, as in the case of Lie groups.

[II] For example, the superposition principle, which is an essential element of the entire formalisms of quantum mechanics and electrodynamics (and drastically simplifies calculations), would no longer be valid.

Part III: The Mathematical Universe

functions that link Lie algebra with the Lie group, for example the generalized exponential and logarithmic functions for matrices. To rediscover their well-known definitions in the more general form of groups is certainly appealing.

> The exponential for matrices is defined by a straightforward analogue to the real numbers, where the following rule holds:
>
> $$e^x = 1 + x + \frac{1}{2}x^2 + \frac{1}{6}x^3 + \frac{1}{24}x^4 + \cdots$$
>
> If we consider the Lie group SO(3), which can be expressed by 3×3 matrices with determinant 1 (see Euler angles), the corresponding Lie algebra is described by the antisymmetric 3×3 matrices, which could of course also be written as vectors. Instead of x in the above formula, we now simply use a matrix. Powers of matrices are expressed by performing the multiplication in a row, e.g. $x^3 = x \cdot x \cdot x$. This creates nonzero diagonal elements, so that a rotation matrix with determinant 1 is obtained:
>
> $$\begin{pmatrix}1 & 0 & 0\\ 0 & 1 & 0\\ 0 & 0 & 1\end{pmatrix} + \begin{pmatrix}0 & 0 & 0\\ 0 & 0 & -x\\ 0 & x & 0\end{pmatrix} - \frac{1}{2}\begin{pmatrix}0 & 0 & 0\\ 0 & 0 & -x\\ 0 & x & 0\end{pmatrix}\begin{pmatrix}0 & 0 & 0\\ 0 & 0 & -x\\ 0 & x & 0\end{pmatrix} +$$
>
> $$\frac{1}{6}\begin{pmatrix}0 & 0 & 0\\ 0 & 0 & -x\\ 0 & x & 0\end{pmatrix}\begin{pmatrix}0 & 0 & 0\\ 0 & 0 & -x\\ 0 & x & 0\end{pmatrix}\begin{pmatrix}0 & 0 & 0\\ 0 & 0 & -x\\ 0 & x & 0\end{pmatrix} - \cdots = \begin{pmatrix}1 & 0 & 0\\ 0 & \cos x & -\sin x\\ 0 & \sin x & \cos x\end{pmatrix}$$
>
> Here, the Taylor series, a known tool in calculus was used:
>
> $$\sin x = x - \frac{1}{6}x^3 + \frac{1}{120}x^5 \ldots \text{ and } \cos x = 1 - \frac{1}{2}x^2 + \frac{1}{24}x^4 \ldots$$
>
> Analogous formulas hold if one considers complex-valued 2×2 matrices with determinant 1, which, as already mentioned, can also serve as a representation of S^3. The associated Lie algebra then consists of the traceless[1] complex 2×2 matrices.

---

[1] This means that the sum of the diagonal elements is zero.

11 The Three-dimensional Unit Sphere – Full of Surprises

If we think about fundamental physics, the interesting interplay between the Lie algebra and the Lie group could lead to the emergence of a physical constant. Lie algebra and Lie group are qualitatively different objects. If, however, they represent physical quantities contained in a known law of nature, these quantities are usually related in an equation. But then, the qualitative difference between a Lie group and a Lie algebra must be taken into account by a 'physical' constant, which cannot be a pure number. Physical constants could thus arise when we find approximate matches between elements of a Lie group and elements of its algebra, whose true cause we have not yet understood. However, the appearance of such 'constants' would be a mathematical necessity. All this is certainly speculative, but it is clear that further mathematical research is needed to unfold the full potential of the differentiable groups for describing nature.

Part III: The Mathematical Universe

12 How S³ Manifests in Reality

Niels Bohr's insight that electrons orbiting the atomic nucleus have an angular momentum which is always a multiple of the constant $\hbar$ was already spectacular. In 1922, during the subsequent heyday of atomic physics, Otto Stern and Walter Gerlach carried out an experiment that was to elicit an even more puzzling property from nature. Until then, electrons were believed to have not only an orbital angular momentum, but also a kind of self-rotation called spin, which creates a magnetic moment.

Stern and Gerlach sent a beam of silver atoms (they have single electrons in their outermost shell) through an external magnetic field that was supposed to exert a deflecting force on the magnetic moments of the electrons. If the axes of rotation of the electrons were randomly orientated in space, which Stern and Gerlach had presumed, then a continuous distribution of the points of impact was to be expected. To their great surprise, however, the electrons behaved quite differently: they only arrived at two distinct points, as if their spins were aligned exactly parallel or exactly opposite to the magnetic field. In terms of classical understanding, this was absurd, because they seemed to have either been 'preselected' or must have had an incomprehensibly long time to adjust their direction.

> *But the most interesting thing at present is the experiment by Stern and Gerlach. (...) the alignment should take more than 100 years.* – Albert Einstein

LOOKS LIKE ROTATION, BUT IT'S NOT

Apparently, however, something fundamental was wrong: electrons simply cannot rotate in a classical way. It is often said that this is a 'quantum mechanical' effect, whatever that means.

Part III: The Mathematical Universe

Yet, it has little to do with the other mysteries of quantum mechanics such as randomness, wave-particle duality, etc. The quantization of spin, as one might call Stern and Gerlach's result, is an additional, very irritating property of nature. The result of the famous experiment can only be interpreted in such a way that elementary particles basically appear in two different states as soon as they are somewhere located in space. Similarly, the periodic table of the chemical elements can be explained only by assuming that two electrons with opposite spin 'live' in one orbital, which is actually a quite arbitrary rule.

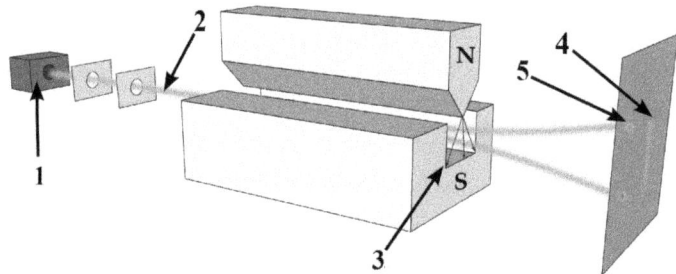

Schematic illustration of the experiment of Stern and Gerlach in 1922: Silver atoms travelling through an inhomogeneous magnetic field, and being deflected up or down depending on their spin; (1) furnace, (2) beam of silver atoms, (3) inhomogeneous magnetic field, (4) classically expected result, (5) observed result, which is completely incomprehensible from a classical point of view.

Nature thus surprises us with a mysterious doubling of states as soon as we consider orientations in space. From the conventional point of view that assumes that it is in space and time that reality takes place, there is no reason whatsoever for this doubling. Apparently, however, the three-dimensional unit sphere S^3 shows exactly this structure. It is *almost* identical to the rotations in three-dimensional space $\mathbb{R}^3$, which makes it understandable that we perceive $\mathbb{R}^3$ as reality. The difference from SO(3) is the double cover, i.e. S^3 provides exactly *two* states for each point in SO(3). It will certainly be difficult to work out this connection in a formally precise way, but it is clear that the double cover must be the deeper mathematical cause of the spin phenomenon, for which there is no explanation in the conventional paradigm of space and time.

THE DIRAC EQUATION: EVERYTHING SOLVED?

In the literature it is often claimed that spin is a consequence of the Dirac equation, an attempt to unify quantum mechanics and relativity worked out by Paul Dirac in 1928. Although this is not the case, I would like to comment on it briefly. To begin with, it is clear that Dirac did not really unify those two theories, otherwise the still missing unification would not be generally seen as the big unsolved mystery of physics. However, what Dirac tried to do was to generalize Schrödinger's equation of 1925 in a relativistic manner.

> Dirac considered the Schrödinger equation $\frac{-\hbar^2}{2m} \Delta \psi = E\psi$ in which the momentum operator $p = i\hbar\nabla$ appears as a square. Schrödinger started out from the simple formula for kinetic energy, $E = \frac{1}{2}mv^2$, which he rewrote as $E = \frac{p^2}{2m}$. However, the term for the kinetic energy $\frac{1}{2}mv^2$ naturally appears in the theory of special relativity[1]. It is an approximation which is added to the rest energy $E^0 = m_0 c^2$ to obtain the total energy E, to which $E^2 = E_0^2 + p^2 c^2$ applies. By setting $E = \sqrt{E_0^2 - \hbar^2 c^2 \nabla^2}$, Dirac had found an unconventional solution and then constructed an algebra that fulfilled the otherwise intractable equation.

However, the relativistic formula for E is much more elementary than $\frac{1}{2}mv^2$ because it incorporates the rest mass. If Dirac had actually found an operator for the rest mass of a particle, he

---

[1] Cf. Chapter 8.

Part III: The Mathematical Universe

would probably have solved the riddle of the origin of mass in general. This, however, cannot be done without taking into account the distribution of masses in the universe, which was unknown at that time (ironically, these cosmological considerations on Mach's principle, which Dirac first took up in 1937, are not appreciated at all today).

Because of the problem of mass, Dirac's idea that the Schrödinger equation could simply be generalized could not work in 1928 and his attempt, though ingenious, was actually doomed to fail from the outset. Dirac's declared goal of calculating the rest mass of the electron was not achieved either. Instead, the resulting equation has solutions that correspond to an electron with negative mass. This is clearly unphysical and cannot be regarded as a prediction of the antiparticle of the electron (the positron), as it is often claimed; for the positron has positive mass and energy. What changes instead is the sign of the electric charge – but this does not enter Dirac' calculation anywhere. To consider the Dirac equation as a theoretical rationale for the positron is thus merely a semantic reinterpretation of a failure, which was incidentally Dirac's own assessment of his efforts. He wrote: [55]

> *I really spend my life trying to find better equations for quantum electrodynamics, so far without success.*

A certain amount of legend has also arisen with regard to the 'explanation' of spin.[1] Dirac had reformulated the above expression for the total energy E and then was trying to find an operator whose square resulted in the expression under the root. With undeniable creativity, he found out that this was only conceivable

---

[1] I will not address here the mysterious nature of the distinction 'fermions' (half integer spin) and 'bosons' (integer spin) for particles, since the origin of spin must be understood in first place. These terms are probably not helpful for solving that problem.

12 How S3 Manifests in Reality

by extending the previous arithmetic operations, giving up on commutativity. The resulting Dirac algebra, named after him, describes electrons with a wave function that consists of four (!) complex-valued components, a so-called spinor – not exactly a paragon of simplicity. Above all, however, noncommutativity occurs in a whole series of algebras – not least the quaternions.

The handling of non-commutative algebras, and even the tricky extraction of the square root from the Laplace operator Δ, had actually already been done by William Hamilton; and one can therefore speak of a 'rediscovery' on the part of Dirac.[56] All in all, there are many indications that electrons, including their strange spin behavior, are described more simply by S^3. In any case, despite the elegant representation Dirac had developed, it cannot be claimed that this sheds light on the reason for the *existence of* spin.[1]

EINSTEIN'S DRY LOGIC AGAINST QUANTUM THEORY

Incidentally, the experiment of Stern and Gerlach described above was only one of many examples that contradicted classical concepts. Just remember the thought experiment of Einstein, Podolsky and Rosen (EPR) discussed in Chapter 9. According to conventional wisdom that retains space and time as a base, a system of two electrons with opposite spin can be spatially separated. The problem was that despite the random outcome of one spin's measurement, how could the electron tell the other spin *instantly* to align in the opposite direction? Einstein argued that this could not be the case, because information would then have to propagate faster than light – according to his logic, which was

---

[1] The spin matrices introduced in by Pauli 1927 are also equivalent to the unit quaternions. This also suggests a physical relationship.

Part III: The Mathematical Universe

based on space and time. He was to be proven wrong, though not because nature contradicted relativity, but nature contradicted that kind of logic. Anyway, for a long time the experiment was considered impracticable.

It was only after Irish physicist John Stewart Bell had put Einstein's thoughts into a set of inequalities that were more accessible to experiment that Alan Aspect was able to demonstrate this 'nonlocal' behavior of nature in the 1980s. The innocent term 'nonlocality' means that our image of reality is shattered to the core. Whichever way you look at it, the result contrasts the conventional logic of how processes occur in space and time, and is sometimes referred to as 'spooky action at a distance'. Of course, the cause in this particular case cannot be superluminal velocity, but our idea that two spatially separated electrons align their spin according to mutual communication must be fundamentally wrong.

Rather, two electrons probably form an extended entity which, for compelling reasons, occupies two opposite states, similar to the unit quaternions (a, b, c, d) and (a, -b, -c, -d) that correspond to the same three-dimensional rotation. Although the explanation of the EPR paradox by means of S^3 is certainly not settled, it is clear that the double cover of SO(3) must be the mathematical cause of this weird double occurrence of particles. Since more than 80 years of headaches for theoretical physicists have not brought about a satisfactory solution as long as the concepts of space and time were maintained, this must be the crux of the matter.

It is probably not sufficient either to modify the conventional spatiotemporal wave function by replacing complex numbers with unit quaternions, i.e. to assume a function $(\mathbb{R}^3, \Lambda) \to S^3$. We may also be compelled to replace the usual space $\mathbb{R}^3$, which we perceive as surrounding reality, with S^3. Extremely interesting

12 How S3 Manifests in Reality

ideas towards such a paradigm shift were recently published by the British mathematician Joy Christian.[57]

According to Christian, replacing $\mathbb{R}^3$ by S^3 has similar consequences to the transition from a 'conventional' two-dimensional world to a Möbius strip.[1] Apparently, every object on the Möbius strip would then exist in two mirror-inverted states, corresponding to the 'front' and 'back', although there is of course only one side. If you move to a point at a spatial distance of half the band length, you will find yourself at the same position, only with an 'inverted' perspective. According to Christian, the paradoxes of the Einstein–Podolsky–Rosen thought experiment are just as illusory as those of the Möbius strip. The twisting of the paths is extrinsic on the Möbius strip, while S^3 has an intrinsic twist related to the Hopf fibration.

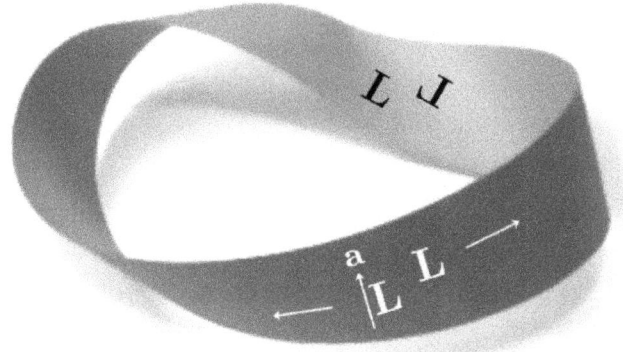

Möbius strip. If one realizes that 'front' and 'back' correspond to only one position in our three-dimensional reality, one may say that every object exists also in its mirror-symmetric state.

Christian's work has unfortunately led to a technical controversy over John Bell's theorem.[58] The revolutionary implications

---

[1] The standard example of a surface which is not orientable. You can easily make a Möbius strip yourself by cutting and twisting the ends of a paper strip by 180° and gluing them together.

for space and time, however, do not seem to have been understood by his colleagues. In a similar vein, reflections about spacetime were made by the Irish mathematician Brian O'Sullivan, who also believes that the Hopf map plays a major role.[59]

WHERE DOES THE UNCERTAINTY PRINCIPLE COME FROM?

In addition to these relatively modern developments, there are the old puzzles of quantum mechanics that have been causing unease since their discovery, but have so far been described within the paradigm of space and time. Particularly noteworthy is Werner Heisenberg's uncertainty principle, which states that pairs of certain physical quantities such as position and momentum, energy and time or even different directions of angular momentum cannot be measured simultaneously. Of course, there is not the slightest reason why this should be a consequence of Newtonian mechanics. But in contrast to spin, the mathematical cause cannot be the double cover of SO(3) by S^3. We need yet another reason for the existence of the fundamental constant h.

How can h emerge from pure mathematics? The mathematical formalism of Heisenberg's uncertainty relation gives us a hint. It is said that when performing a measurement, the commutation of the corresponding operators (for example position and momentum) results in the constant h, while according to classical ideas, the order of measurement should not matter at all. Obviously, however, the very characteristic feature of S^3 is precisely that when multiplying its elements, the order of the factors plays a role. Quantitatively, this is expressed by the term (a b a^{-1} b^{-1}),[1] the so-called commutator. If a and b are themselves small numbers,

---

[1] For conventional numbers, this would be 3·5·1/3·1/5, with the result 1.

12 How S3 Manifests in Reality

which one must assume for any real-world application,[1] the commutator produces values that are again much smaller than the two factors a and b. This reminds one of the tiny value of the quantum of action h=6.626·10^{-34} kg m²/s. So if we look for a purely mathematical cause for the occurrence of the fundamental constant h, the most obvious candidate is the noncommutativity of the multiplication in S^3. Of course, there is no way to understand this characteristic of S^3 within 'straight' Euclidean spaces intuitively, and its artificial integration into conventional space-time has therefore created rather cumbersome formalisms. One of them is the commutator for operators such as the angular momentum. The smallness of the outcome suggests subtle physical effects that have not remained undiscovered for centuries by chance.

> In quantum mechanics, the commutation relations for the angular momentum operators in different directions, L_x, L_y and L_z, are $[L_x,L_y]=[X_yP_z-X_zP_y,X_zP_x-X_xP_z] = i\hbar L_z$, $[L_y,L_z] = i\hbar L_x$ and $[L_z,L_x]=i\hbar L_y$, where X represents the respective position operator and P the respective linear momentum operator. Here it becomes very well visible that noncommutativity is related to the existence of the constant h. The linear momentum operators, for example, are defined by $P_x=i\hbar\frac{\partial}{\partial x}$. They are proportional to spatial derivatives, which cannot be interchanged with the position operator X. This can even be understood by elementary calculus (primed ' indicating the derivative): (x sin x)' = x·cos x + sin x is not the same as x·(sin x)' = x·cos x. It makes a difference whether you first multiply by x and then take the derivative or vice versa.

---

[1] Otherwise, the noncommutativity would certainly have been more conspicuous in experiments. It is only when considering small elements that S^3 can look approximately like space-time.

Part III: The Mathematical Universe

Here again, the fact that the connection between h and non-commutativity is intuitively quite evident should not hide the reality that there will be huge difficulties to be overcome before the concepts can be identified in a satisfactory manner. However, what is conventionally called the measurement of physical quantities must have to do with multiplication of S^3, whose elements presumably represent the reality around us.

> *Quaternions do not commute. Spin also has this property. This bug seems like a feature to me.* - Doug Sweetser

h, c AND THE ORIGIN OF MATTER AND LIGHT

It is striking that the occurrence of the constant h is always related to properties of matter. One may object that Einstein's formula for the photoelectric effect $E = hf$ concerns only light and no matter. However, all measurements of the light quanta's energies are obtained through interaction with matter. When searching for the deeper meaning of the constants h and c, it is clear that h has to do with matter, and c with light.

Given that the properties of S^3 seem to be the cause of the existence of h, we now turn our attention to the equally fundamental constant c. In contrast to the appearance of h in the microscopically small world of atoms, the effects of the speed of light c usually become visible on large scales, e.g. in astronomy or even cosmology.

We come to know about all astrophysical events after a long time delay. Because of the travel time, the sunlight that warms us in the morning takes about eight minutes to reach us, and some supernova explosions have happened billions of years before we detect their light. Conversely, distant civilizations will receive signals from our planet with corresponding delay (maybe this is not a bad thing), if we ever manage to communicate our existence

12 How S3 Manifests in Reality

at all. One can therefore imagine that at any given moment, a spherical shell of information is hurtling out into space, the entirety of which is called the light cone. Accordingly, the light cone of the past is represented by all those points in space-time whose light reaches us at a given point in time. The light cone, of course, fills the entire three-dimensional space, but there is only one specific instant of time at each location that defines the affiliation to the cone.

> In the context of Minkowski space-time, this fact is described by a metric $ds^2 = c^2 dt - dx^2 - dy^2 - dz^2$. The distance s is positive if one remains in the same point in space (dx=dy=dz=0), while purely spatial distances at simultaneity (dt=0) are counted as negative distances. ds=0, on the other hand, denotes those points in space-time that are connected by a light signal and thus defines a light cone.

ON THE LIGHT CONE

For a more visual understanding, let us consider one dimension less and imagine a two-dimensional space (e.g. a water surface) to which we attach time as a third dimension. Then, for example, a water wave that spreads concentrically from a point can be described by a cone in this 2+1-dimensional 'space-time'. Accordingly, we may speak of a light cone in 3+1-dimensional space-time when contemplating the concentric propagation of a light wave out into the universe.

From the perspective of natural philosophy, the view just mentioned is not justified by anything. For there is no reason why the velocity of propagation should not exceed a certain limit, namely the arbitrary constant c. Therefore, we must look for the cause of the existence of c in mathematics.

Part III: The Mathematical Universe

Strictly speaking, we detect light signals only in the 'here and now', while deducing their origin from models. If one wants to question the structure of space and time, the conventional image of the light cone needs to be expressed in a different way. A logical possibility would be to regard the light cone as a horizon space at a certain point of S^3. It would thus represent the collected information that comes together in the 'here and now', originating from the entire universe, which is a three-dimensional manifold even in conventional terms.

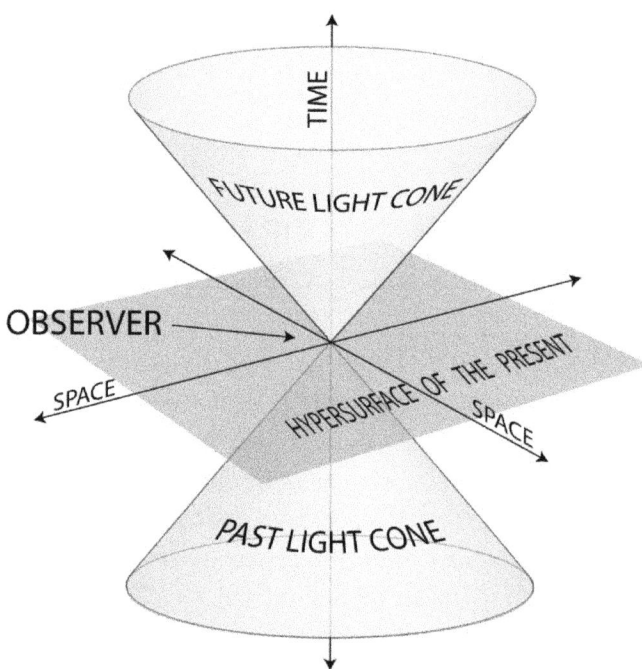

Schematic representation of the light cone. Temporal distances from the origin are counted positive, spatial ones negative, for the light cone itself ds=0 holds.

Incidentally, the horizon space, like a plane attached to a sphere, is only an approximation for S^3, which deviates more and more from it with increasing distance. This means that as a matter

12 How S3 Manifests in Reality

of principle, we would be unable perceive reality exactly. In the conventional model, we receive all information through light which spreads out in straight lines. Analogously, the flat horizon space into which we look from a point on S^3 would be a fundamental limitation for the perception of a curved world, much like the horizon that limits a navigator's view of the sea. This would entail epistemological consequences as to how reality can be perceived at all over great distances. In any case, the idealized Euclidean[1] space $\mathbb{R}^3$, which we seem to perceive by means of light signals, should not be taken as an image of the universe we can trust. Instead, one should be prepared for a situation where, with increasing distance, reality could deviate considerably from this impression. In the conventional model, there are spatial distances that can be converted into temporal distances when divided by the speed of light c. If we give up the concepts of space and time, the definition of distance does not, however, lose its meaning: distances can still be specified in the horizon space of S^3.

The most striking observation at cosmological distances is the redshift of light from distant galaxies. I have outlined previously how redshift can be explained within a variable speed of light model without having to resort to auxiliary assumptions such as an expansion of the universe. Nevertheless, this explanation presupposed the paradigm of space and time. In a model of the universe based on S^3, observations of distant galaxies would be distant points in the horizon space, which is different from the 'real' S^3, which is curved. One may wonder whether the cosmic redshift is merely the distortion of reality that would necessarily occur if we received information from the horizon space instead of from distant points on S^3. Nevertheless, this remains speculation.

---

[1] In conventional general relativity, space is believed to be curved, which has little to do with my argument here, however.

Part III: The Mathematical Universe

It is interesting that as early as 1917 – before redshift and galaxies had even been discovered – Albert Einstein was thinking about whether the cosmos as a whole could be described by S^3, although many of its properties were unknown to him.[60]

IS LIGHT THE LIE ALGEBRA?

The horizon space is a natural concept when considering curved manifolds like the S^3. If one associates it with the light cone, the question arises as to what physical meaning the tangent space, i.e. Lie algebra, could have. This algebra is a natural characteristic of a differentiable manifold as well. While the directions in the tangent space correspond to those in the horizon space, the radial component in the former has a qualitatively different meaning: intuitively speaking, how fast a rotation is spinning in a given direction. An obvious association here is the frequency or wave number[1] of light – the only parameter in which light waves incident from a certain direction differ.

If one considers light as a phenomenon of Lie algebra, certain difficulties of interpretation arise. Light is usually described as a spatiotemporal electromagnetic wave. However, there is an interesting property of light for which there is no convincing conventional explanation. Light always transmits angular momentum as soon as it emits a quantum of energy (I keep using this term, although it is somewhat misleading). Again, there is no a-priori law of nature prohibiting light from simply transmitting energy without any angular momentum. Light seems to carry a rotation within itself, another property nobody has been able to explain so far.

If in the S^3 model light is considered a consequence of the fact that tangent vectors, i.e. elements of Lie algebra so(3), may exist

---

[1] The reciprocal of the wavelength.

12 How S3 Manifests in Reality

everywhere, then these vectors necessarily have something to do with rotations. From a traditional point of view, light is something that propagates in straight lines, but intrinsically rotates for unknown reasons. If, on the other hand, light is regarded as a phenomenon of the tangent space defined by the Lie algebra of S^3, it would be something completely straight that manifests in a space that is inherently twisted. Perhaps our perception is deceived in exactly this way as soon as we, like Newton, postulate a Euclidean space as reality.

MATHEMATICS PRODUCES TWISTS

The phenomenon of the 'contortion' of a space deserves to be examined more closely. If we recall the complex numbers in the two-dimensional plane, it is noteworthy that it was only the aspiration to define multiplication in a reasonable manner that led to the occurrence of rotations in the complex plane. All two-dimensional rotations can therefore be represented by multiplication with complex numbers of unit length, which correspond to S^1. Defining multiplication in higher dimensions requires at least four, and additionally, quaternion algebra (Hamilton had a hard time figuring that out).

As in the case of complex numbers, however, rotations can be described with one dimension less, namely by the unit quaternions or S^3. The rotations that now become possible are not only an extension to three spatial directions, but add a new quality to this type of multiplication, which is best imagined as twisting, shearing, or screwing.[1] Originally, there were four directions in space (a, b, c, d), and a simple rotation involves two directions. Hence, there are three ways to couple two pairs of rotations (e.g. ac-bd),

---

[1] It should be noted that this goes beyond the already mentioned representation of three-dimensional rotations, because in that case, the 'screwing' part of the complex rotation was neutralized by the conjugates **q** and **q**$^{-1}$ (cf. Chap.11).

Part III: The Mathematical Universe

each rotation having a certain orientation.[I] Here again, pure mathematics creates a surprising complexity!

If you try to wring dry a towel with your hands, you can do this in two different ways, by turning *both* ends either clockwise or anticlockwise. This introduces screw sense, which does not change, incidentally, if you chose a different axis of rotation in space. In physics, this is a well-known property of elementary particles and called helicity.[II] Needless to say, no one had predicted this phenomenon, and again, there is no theoretical reason why it should have occurred in nature. If one looks for an explanation from first principles, it seems a good idea to consider the properties of multiplications in S^3. If this is the right track, then helicity would turn out to be an inevitable consequence of multiplications in three dimensions.

In a skew field like quaternions, the collapse of commutativity with $p \cdot q \neq q \cdot p$ is inevitable, which leads to a necessary distinction between multiplication from the left and from the right. This, too, is an interesting complexity that emerges only in higher dimensions. One may speculate about a connection to electric charges of different signs, for whose existence physics could not give a reason so far. Although this association seems rather vague, the natural-philosophical approach must nevertheless search for causes. This holds also for the obvious conceptual difference between gravity and electricity.[III] Traditionally, electrodynamics is described by vectors fields such as the electric and

---

[I] Hopf fibration is also related to this: due to its inherent twist, S^3 cannot be simply decomposed into S^2 and S^1. For visualization, I again recommend the videos of Ben Eater.

[II] This was shown, among other experiments, by the decay of the cobalt-60 nucleus, for whose interpretation the Nobel Prize was awarded in 1957.

[III] As the introductory chapters have certainly made clear, I do not consider the concepts of the so-called strong and weak interactions to be useful.

magnetic fields E and B. But if these fields are actually quaternions (this is speculation), which only *look* like vectors, one would expect subtle effects in strong fields due to the non-commutativity of rotations.[I]

WHY THREE DIMENSIONS?

Since the (four-dimensional) quaternions themselves already exhibit very interesting properties, one could ask why we should limit the discussion to unit quaternions. Perhaps one could relate the four dimensions expressed in the coordinates (a, b, c, d) of a quaternion with the 3+1 dimensions of space-time. However, this does not explain the *qualitative* difference between space and time, which are phenomenologically so disparate. Such an essential difference should result from the structure of quaternions. There is such a natural 3+1 separation in quaternion algebra, yet it is already found in the unit quaternions or S^3, to which the restrictive condition $a^2+b^2+c^2+d^2=1$ applies.

Once such an algebra is introduced, the four elements a, b, c and d are not of the same kind, because only *a* represents a real number, while b, c and d are the components of the complex units i, j and k (in the analogous case of a complex number a+bi, *a* was the real part). This becomes clear when the product is executed.[II] Thus even in S^3, a three-dimensional manifold, a system of four components naturally emerges. The three coordinates (b, c, d) are analogous to the imaginary part of the complex numbers, which play a major role in quantum mechanics. However, it is hardly conceivable that the abstract dimensions of the wave function can be assigned to real space.

[I] This is elaborated in my article vixra.org/abs/1901.0083, originally motivated by MacCullagh's aether theory from 1839.
[II] See the algebraic rules in Chapter 10.

Part III: The Mathematical Universe

As far as quaternions are concerned, a direct identification, such as the real number a as time component, would certainly be mistaken. It is probably more promising to write each quaternion as the product of a unit quaternion (of unit length) and its actual length (the norm), as Hamilton had already done.[1] It would not be out of the question to speculate whether the norm (length) of a quaternion could represent time. In that case, space-time would be a continuously growing quaternion.

In view of the unsettled nature of these considerations, it is difficult to decide whether the general four-dimensional quaternions can provide a description of time or whether S^3 alone plays a fundamental role and time turns out to be an emergent phenomenon. Both are hypothetical approaches, which may also lead to contradictions. Nevertheless, the chances of explaining the phenomenon of time are not so bad, given the intriguing algebraic relationships we have observed.

THE END OF CONSTANTS OF NATURE

Let us try a summary here. Just as the fundamental constant h seems to originate from the noncommutativity of S^3, the constant of nature c could result from the fact that S^3 has a tangent space. One may rightly regard these or the preceding remarks as speculative, but from the point of view of natural philosophy there is no other way than to try to link the cause of the existence of these fundamental constants to mathematical properties. If one takes the reverse perspective, this becomes perhaps more obvious: what special properties does S^3 possess that go beyond the conventional computational laws of physics?

---

[1] Hamilton described the unit quaternion as *versor*, and the norm as *tensor*, which might be confusing terms in today's usage.

12 How S3 Manifests in Reality

No one can deny that tangent space and noncommutativity are two salient characteristics that can create the illusion of the constants c and h. Even the natural phenomena of turns and twists can only find a real justification in the mathematical structure of S^3. The probability that the properties of S^3 are related to h and c is therefore very high. On the other hand, there is certainly a long way to go before a consistent formalism can be developed that finally clarifies these relationships. However, this is not surprising, given the enormous conceptual difficulties that arise when we bid farewell to space and time and try to describe reality with a mere three-dimensional object.

Part III: The Mathematical Universe

13 Unsolved and Crazy Stuff and Pure Mathematics

There are excellent reasons to assume that the three-dimensional unit sphere S^3 provides an explanation for the existence of elementary constants of nature. Nevertheless, some serious problems should be mentioned here, especially those concerning the interpretation of the horizon. If one identifies all light signals from the universe that meet in the present, i.e. at a given point in space-time, with the horizon space of a single point of S^3, the consequences go beyond imagination. How to understand the passage of time we are accustomed to? The fact that we perceive the world around us like a movie can be imagined as moving along a path on S^3, which creates a sequence of (Euclidean) horizon spaces. On the other hand, time is anything but a path in S^3. Such a path could follow an arbitrary trajectory and even return to its origin, while time seems to flow inexorably in one direction.

One solution to the problem could be to recognize only the present, the 'here and now', as reality. All the past would not exist in the proper sense. There would be only images from distant parts of the horizon space, which create the illusion of the past, while the corresponding events took place at a distance corresponding to the light travel time. All other manifestations of the past, in particular memories, notes, and any form of recording, would only be tools with which information from the horizon space is transformed back into the present. Everything we imagine as past – i.e. all historical, geological and cosmological events – would then be mere illusions created by the false paradigm of a 3+1-dimensional space-time. Only the present would remain real.

But even with such an admittedly exotic interpretation, the phenomenon of the passage of time would still not be clarified.

Part III: The Mathematical Universe

Why does time run so relentlessly in one direction? Why can't we stop it? This remains perhaps the greatest mystery when applying S^3 to the problem of space-time. It is probably wise to first thoroughly examine all the surprising mathematical properties of S^3 before trying to identify them with physical phenomena. Interesting concepts are the geometrical 'Flow' or 'Ricci Flow' with which Grigori Perelman had proved Poincaré's conjecture. Whether such flows can ever explain the phenomenon of time, however, remains questionable.

> *The distinction between past, present, and future is only a stubbornly persistent illusion.* — Albert Einstein, 1955

NO MORE EQUATIONS POSSIBLE?

If we associate, as proposed in the previous chapter, the Lie algebra so(3) of S^3 with light, we must clarify the meaning of the vectors, i.e. the elements of the Lie algebra. The direction remains the conventional one in space. However, the length of the vectors that we have related to light can only correspond to the parameter that characterizes light, namely frequency. This would mean that all information we receive about reality would be described by Lie algebra vectors. However, it remains unclear how the apparently continuous flow of information we are exposed to is expressed by means of S^3 or its Lie algebra.

Progress can only be made by further analysis of the mathematical structures of S^3. First of all, one has to deal with the difference between the Lie group S^3 and the Lie algebra so(3), two notions that are not very intuitive. Therefore, instead of S^3, we imagine again SO(3), a Lie group that has the same Lie algebra as S^3, namely so(3). But so(3) and SO(3) are not to be confused! The latter describes global rotations, so(3) only their rate of change. SO(3) cannot become arbitrarily large, because after a full turn of 2π (in case of the S^3, 4π) one gets back to the starting

13 Unsolved and Crazy Stuff and Pure Mathematics

point. Conversely, the rate of change of the rotations, so(3), can become arbitrarily large, because it is calculated from the quotient $\Delta\varphi/\Delta t$, in which the parameter Δt can be infinitely small. The Lie algebra so(3) can thus be seen as a derivative of the Lie group SO(3) (and, at the same time, of S^3) and the fact that so(3) and S^3 differ so much from each other is a characteristic of the three-dimensional unit sphere.

In physics, there is a large number of equations that attribute functions to their derivatives and therefore are called differential equations. The most prominent examples are Maxwell's equations in electrodynamics, Einstein's equations in general relativity, and the Schrödinger equation in quantum mechanics. The formulation as an equation is justified by the fact that the corresponding objects – i.e. electromagnetic fields, the Einstein tensor or the wave function – do not change their quality during the process of differentiation; in other words, they remain functions of space and time similar in character.[I] But if S^3 were to become the stage of physical reality, then these celebrated theorems of physics would possibly be comparing apples to oranges.[II] For if one defines the corresponding objects on S^3, they can no longer be equated with their derivatives. A Lie algebra is just qualitatively different from a Lie group. If one equates these different objects anyhow, their different quality must be accounted for. Since the Lie algebra is closely related to the concept of the tangent space, one may hope that the fundamental constant c, or rather its still unknown origin, will be able to provide such an explanation in

---

[I] The calculus of differential forms, prevalent in differential geometry, has many advantages and provides further insight. For example, the (outer) derivative of a 2-form results in a 3-form and so on. But even this sophisticated method does not work on S^3.

[II] Most of the common mathematical tools, such as vector and tensor analysis as well as function spaces in quantum mechanics, involve so-called linear operations, which are generally easier to handle. However, there is no way that a nonlinear object such as S^3 could be treated with similar methods.

Part III: The Mathematical Universe

the future. However, this would mean that practically all the theorems of theoretical physics are at best approximations.

Sophus Lie (1842-1899)

THE SPEED OF LIGHT AND THE DERIVATIVE

Since the analysis of S^3 has shown that differentiation can change the character of a function, one may consider the hypothesis that the mere existence of differential calculus leads to the appearance of constants of nature. Evidently, the process of differentiation produces the Lie algebra so(3) from the Lie group S^3. If the former can be identified with light waves, the cause of the constant of nature c or the very existence of light would be a trivial consequence of the mathematical fact that manifolds have derivatives. Assuming this is a worthwhile approach, it would be obvious to identify c with the first derivative. Then of course, one may further speculate whether the constant h is related to the second derivative. More obvious, however, remains the connection

13 Unsolved and Crazy Stuff and Pure Mathematics

of h with noncommutativity, which is only indirectly related to the second derivative or curvature.

Let us return to the consideration of how the conventional bundle of space-time ($\mathbb{R}^3$, Λ) can best be replaced. When describing the physical phenomena, I speculated whether complex-valued wave functions and vector fields would be better replaced by a fiber S^3. However, the arbitrariness in the construction of space-time can only be eliminated if the bundle is replaced as well. Thus, one is led to consider the fiber bundle $S^3 \to S^3$, i.e. a mapping from the three-dimensional unit sphere on itself; however, this raises new mathematical and conceptual difficulties. Is S^3 the only option? Or, for example, would something like a large number of maps $S^1 \to S^3$ be worthwhile to analyze? Or even the S^7?

Unfortunately, it is much easier to convince oneself that the conventional picture is inadequate than to find a priori convincing arguments for an alternative structure of reality. Mathematically, maps like $S^3 \to S^3$ have interesting properties: for example, topological defects are classified by higher homotopy groups (so far we have considered only the first homotopy group). The properties of such topological defects in many respects resemble those of elementary particles, especially with respect to pair generation and annihilation.[61] However, these parallels have so far been drawn within the conventional space-time picture.

IS THERE A CONSTANT OF NATURE MISSING?

Certainly, the two constants h and c could emerge from the properties of S^3. However, in the discussion about fundamental constants at the end of Chapter 7 it became clear that there were not only two, but three remaining free parameters calling for an explanation. The missing parameter was the epoch τ, a dimensionless number derived from the ratio of the largest and smallest structures, i.e. the radius of the universe and the proton (one can

Part III: The Mathematical Universe

also formulate this ratio by using time, but it makes no difference).

Unfortunately, however, we must also question the constant τ for two reasons. Firstly, the notions of space and time, to which it refers, lose their meaning in a description of reality by means of S^3, and secondly, τ is undeniably a concrete numerical value that is not justified by anything. If we abandon the idea that the world evolves, as time passes, in a three-dimensional space, then the 'explanation' of the epoch τ, which assumes that we are living only 'now' at a certain moment during the evolution of the universe, is no longer satisfactory.

From a conventional point of view, it would be expected that the value of the epoch (which boils down to measuring the Hubble constant), changes slowly. Assuming we can achieve the necessary precision, one would expect a slightly different value a few years later, for example. Since such an accuracy has not yet been reached, a surprise cannot be ruled out. In principle, the epoch could also prove to be a constant (this would be a blow to current beliefs). If one is ambitious enough to want to explain such a constancy, then this huge number 10^{40} would have to be created by pure mathematics. 10^{40} is approximately the measured size of the universe in units of the proton radius. However, as the theory elaborated in Chapter 5 makes clear, this value is a consequence of the apparent shortening of scales, whereby the 'absolute' epoch is 10^{53}. This raises the question whether such a number can arise from pure mathematics at all. It was not for nothing that Paul Dirac, who had been working on a related problem before 1938, had his doubts. He thought the number was just too big to appear in any reasonable mathematical context.

13 Unsolved and Crazy Stuff and Pure Mathematics

CONTINUOUS AND DISCRETE GROUPS

However, there have been important discoveries in mathematics in the meantime, which I would like to report on in this context, although the related speculation I will mention seems almost crazy. We are talking about the classification of so-called finite simple groups that was completed in 2004. Simple group means that it cannot be split up into subgroups without destroying the group structure. For example, vector addition in two-dimensional space is not a simple group because it can be broken down into adding the x- and y-components, respectively. The group of rotations in three-dimensional space, on the other hand, is simple, because it cannot be decomposed into partial rotations around the three spatial axes (cf. noncommutativity and Euler angles in Chapter 11). Analogous arguments apply to discrete groups.

Let us have a look at an example of a simple finite group that is easy to grasp. The symmetry group of a cube consists of rotations and reflections that transform the cube into itself. This group can be effectively illustrated by looking at the rotation axes, each of which passes through the center of opposite faces or edges, or through opposite corners.

If the axis of rotation passes through opposite surfaces (three possibilities, see picture), there are four angles: 0°, 90°, 180°, 270°, i.e. three elements different from the identity. From the initial state, 3·3=9 symmetry operations of the cube can thus be carried out by rotating it around the centers of the surface. If one puts the axis of rotation through the opposite corners (four possibilities), there are 4·2=8 additional rotations 0°, 120°, 240°), and with the axes through the opposite edges, another 6·1=6 rotations, which altogether add up to 24 elements, if one includes identity, the neutral element.[1] This rotating group is simple and a

---

[1] Considering reflections as well would add another 24 to a total of 48 elements.

Part III: The Mathematical Universe

discrete subgroup of SO(3), the rotations in three-dimensional space. Analogously, symmetry groups of other Platonic solids can be determined.[1]

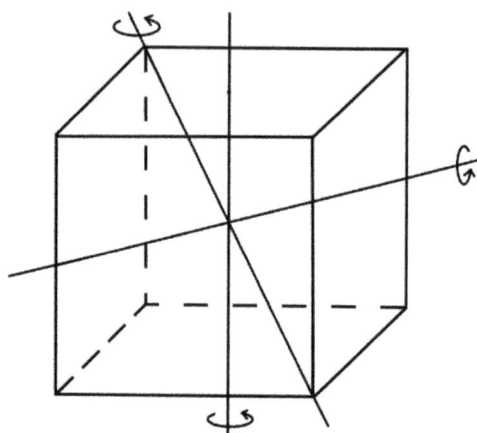

The symmetry group of the cube. For each of the different rotation types, only one generic axis is shown. There are three axes through the surfaces (only top-bottom is shown), around which rotations of 90° leave the cube unchanged. The same applies to the six axes through the edges (rotation of 180°) and the four axes through the corners (120°). Similarly, symmetry groups of other Platonic solids such as a tetrahedron or dodecahedron can be constructed. However, all the resulting rotations are also contained in SO(3).

IS NATURE DISCRETE?

Now imagine the wide field that opens up when considering symmetry groups in higher dimensions, or any other group with a finite number of elements. In a huge collective effort that in-

---

[1] The symmetry group of the octahedron corresponds to that of the cube, that of the icosahedron to that of the dodecahedron, and then there is the tetrahedron. These Platonic solids actually reflect all discrete subsets of SO(3).

13 Unsolved and Crazy Stuff and Pure Mathematics

volved hundreds of publications, mathematicians actually succeeded in identifying all these groups. Perhaps not surprisingly, they included colossal entities, remote from any possible imagination. In 1984, the mathematicians Fischer and Gries discovered the largest group of all, which was named after them – the Fischer-Gries *friendly giant* or, less politely but more commonly, 'monster'. It contains $2^{46} \cdot 3^{20} \cdot 5^9 \cdot 7^6 \cdot 11^2 \cdot 13^3 \cdot 17 \cdot 19 \cdot 23 \cdot 29 \cdot 31 \cdot 41 \cdot 47 \cdot 59 \cdot 71 = 8.08 \cdot 10^{53}$ elements. You see what I'm getting at. This number – in fact the largest number that pure mathematics can give us – is of the same order of magnitude as the absolute epoch discussed in Chapter 5, which was distilled from sophisticated observations of the universe. This coincidence is funny, but nothing more. Unfortunately, there is no mathematical, physical or any other connection in terms of which these numbers could be related to each other; and I have no idea on this topic I would call reasonable. I am merely sharing the coincidence, should any genius ever be able to justify it.

I'd like to know what the hell is going on.–
John Conway

In the present analysis of the history of physics, we have not deviated from the principle of simplicity, i.e. we have not accepted the postulation of any arbitrary numbers. The working hypothesis, backed by historical evidence, was that it should be possible to explain reality by means of mathematical structures that are as simple as possible. The three-dimensional unit sphere proved to be particularly promising, not only because it was sufficiently simple, but above all because its special properties reflected a whole range of physical phenomena.

However, the statement that the S^3 is simple (in a methodological sense) could be objected to from a very general point of view. After all, S^3 consists of a continuum of infinitely many elements, ultimately even more numbers than there are particles in the universe, for example. A puritan logician might argue that we are using unnecessarily many numbers to describe nature and – so the possible criticism may argue – it is easy to praise the alleged simplicity of S^3 if previously the number system had been inflated by postulating a continuum with infinitely many elements.

After having invented the natural numbers $\mathbb{N}$ and the integers $\mathbb{Z}$, mankind very soon used (infinitely many) rational numbers $\mathbb{Q}$ before finally, by including roots and transcendental numbers like π, the number space was extended to the continuum of real numbers $\mathbb{R}$. Since Newton and Leibniz used $\mathbb{R}$ for differential calculus, a continuum of numbers has proved to be an incredibly useful tool for describing the laws of nature.

Nevertheless, one may still raise the philosophical question whether a continuum of numbers is really indispensable for the description of reality with its finitely many, or at best countably infinite, phenomena. If one were to abandon the continuum and all constructions based on it – which would mean sacrificing the greater part of 400 years of mathematics – one would, however, have to return to the theory of groups, among which the simple finite groups play a special role.

From this broader perspective, one could therefore speculate that simple groups have some hitherto unrecognized relevance for fundamental physics. At the same time, the strangeness of this branch of mathematics would give a bleak perspective on everything we believe to be physical reality, which is why I prefer a number continuum despite the above philosophical objection.

Outlook

After this long journey through the history of physics, I hope I have convinced you that the long-term goal of theoretical physics must be to do without any constants of nature. As in the case of belief in gods, postulating arbitrary numbers has never advanced physics. However, this insight into 'constants of nature' not only shows that what was believed to be basic research since about 1930 must be considered obsolete from a fundamental perspective, but it also forces us to recognize that the constants h and c are anomalies in an epistemological sense. This has the inevitable consequence that the concepts of space and time are of limited usefulness when describing reality properly, a fact that unfortunately affects the whole of the natural sciences since Newton.

Questioning the concepts of space and time has serious 'side effects' on all sciences. We would have to suspend practically all fundamental insights of the last 400 years and be prepared for abysses or new perspectives to open up. For example, the following notions would have to be redefined if were no longer to recognize space and time as fundamental: causality, randomness, determinism, evolution. They all make statements about spatio-temporal patterns, which we would at least have to question. Of course, space and time will nevertheless remain useful phenomenological terms for everyday life.

On the one hand, our knowledge about nature seems to have suffered a serious setback with these considerations; on the other hand, a new perspective to understand reality on a completely new level has opened up with the three-dimensional unit sphere. Even though many of the connections shown are far from being

Part III: The Mathematical Universe

formally rigorous, I hope that this book will be an incentive, particularly for mathematicians, to use their skills for a truly relevant description of nature.

HELP FROM ARTIFICIAL INTELLIGENCE?

Of course, doubts also remain as to whether we – with our limited abilities of the species *Homo sapiens* – will be at all capable of mastering the Herculean task of transcending space and time and arriving at deeper concepts that bring our understanding of nature to a new level.

Right now, however, humankind is on the verge of deciphering the functions of the brain and is trying to simulate its powerful algorithms on computers – exceeding by a huge factor the brain's speed and storage capacity. This will not only revolutionize science in a conventional way (as computers have already done with regard to data evaluation), but will potentially lead to intelligent machines that will extend the limits of knowledge currently marked by the 'constants of nature'.

So, if those concepts that will eventually replace space and time are beyond our human imagination, there is hope that one day such insights will be gained on this indirect path. At least there would be a starting point for artificial intelligence as regards where to look for a new description of reality.

Acknowledgements

The time to write a book is often taken from family life, and I am grateful again for the understanding I have experienced from my family. Many friends and colleagues have contributed to bringing these thoughts into a more mature form, and I would like to underline the help of my collaborator Jan Preuss. I also feel a special obligation to my country, which has widely been able to preserve individual freedom, the rule of law, and an open, knowledge-based society that has allowed me to devote a large part of my life to the study of the laws of nature. May it remain so.

Literature

Assis, A.K.T., Weber's Electrodynamics, Springer 1994

Assis, A.K.T., Relational Mechanics, Apeiron 2014

Ball, W.W. Rouse, An Essay on Newton's principia, Macmillan 1893

Barbour, Julian: The End of Time, Oxford Univ. Press 1999

Barbour, Julian: The Discovery of Dynamics, Oxford Univ. Press 2001

Bell, John: Speakable and Unspeakable in Quantum Mechanics, Cambridge Univ. Press 1987

Conway, John: On Quaternions and Octonions, Transatlantic Publishers 2001

Einstein, Albert: My world view (German), Ullstein, 1988

Familton, Johannes C: Quaternions, arxiv.org/abs/1504.04885

Heisenberg, Werner: The part and the whole, Piper 1969

Hestenes, David: Space-Time Algebra, Birkhäuser 1966

Hossenfelder, Sabine: Lost in Math, Basic Books 2018

Jammer, Max: Concepts of Space, Harvard Univ. Press, 1954

Jordan, Pascual: Schwerkraft und Weltall, Vieweg 1955

Kragh, Helge: Higher Speculations, Oxford University Press 2011.

Kragh, Helge: Dirac, Cambridge University Press 1990

Kuhn, Thomas: The structure of scientific revolutions, Univ. of Chicago Press, 1962

Kumar, Manjit: Quantum: Einstein, Bohr and the Great Debate About the Nature of Reality, Icon Books 2009

Landau, L.D., Lifschitz, E.M.: Theoretical Physics Volume II

Lindley, David: The End of Physics, Basic Books 1993

Lindley, David: Uncertainty, Anchor Books 2008

Mach, Ernst: The science of Mechanics, Tr. by McCormack, T. https://archive.org/details/in.ernet.dli.2015.154174/page/n353

McCulloch, Michael E.: Physics from the Edge, World Scientific Publishing Company 2014

O'Shea, Donal: The Poincaré Conjecture, Walker Books 2007.

Pais, Abraham: Subtle is the Lord, Oxford Univ. Press 1982

Rosenthal-Schneider, Ilse: T. Braun (ed.). Reality and Scientific Truth: Discussions with Einstein, Von Laue, and Planck; Wayne State Univ. Press, 1980

Sanders, Robert: The Dark Matter Problem, Cambridge Univ. Press 2010

Schrödinger, Erwin: Nature and the Cambridge Univ. Press 1996

Schrödinger, Erwin: My View of the World, Ox Bow Press 1983

Singh, Simon, Big Bang, Harper Perennial 2005

Shamos, Morris H.: Great Experiments in Physics, Dover 1959

Tegmark, Max: The Mathematical Universe, Knopf 2014

Unzicker, Alexander: Bankrupting Physics, Macmillan 2013

Unzicker, Alexander: The Higgs Fake, CreateSpace, 2013

Unzicker, Alexander: Einstein's Lost Key – How we Overlooked the Best Idea of the 20th Century, Create Space 2015

Picture Credits

p.10: upload.wikimedia.org/wikipedia/commons/0/0e/Cassini_apparent.jpg   public domain

p.10: commons.wikimedia.org/wiki/File:Solar_sys8.jpg   public domain NASA

p. 28: en.wikipedia.org/wiki/File:Kepler-first-law-math.svg CC BY-SA 2.0 User:W!B:

p. 28: en.wikipedia.org/wiki/File:Kepler-second-law.svg CC BY-SA 3.0 Arpad Horvath

p. 32: commons.wikimedia.org/wiki/File:Balmer.jpeg public domain

p. 36: en.m.wikipedia.org/wiki/file:Single_electron_orbitals.jpg GNU free documentation license

p.39: commons.wikimedia.org/wiki/File:Hydrogen_transitions.svg CC BY 2.5 User:Szdori

p.42: commons.wikimedia.org/wiki/File:BlackbodySpectrum_lin_150dpi_en.png CC BY-SA 3.0 User:Sch

p.49: commons.wikimedia.org/wiki/File:Einstein1921_by_F_Schmutzer_4.jpg public domain

p. 62: Newtonian Bucket: Author

p. 63: commons.wikimedia.org/wiki/File:Ernst_Mach_01.jpg public domain

p. 89: commons.wikimedia.org/wiki/File:Paul_Dirac,_1933.jpg public domain

S.112: commons.wikimedia.org/wiki/File:Illustration_from_1676_article_on_Ole_R%C3%B8mer%27s_measurement_of_the_speed_of_light.jpg public domain

p.129: commons.wikimedia.org/wiki/File:Erwin_Schr%C3%B6dinger_(1933).jpg public domain

p. 132: commons.wikimedia.org/wiki/File:Schrodingers_cat.svg CC BY-SA 3.0 User:Dhatfield

p. 132: ethz.ch/en/news-und-events/eth-news/news/2015/12/schap-faster-fraction-remove-quantum-dot.html Courtesy of ETH Zurich / Aymeric Delteil

p. 142: commons.wikimedia.org/wiki/File:VectorField.svg public domain

p.144: mathworld.wolfram.com/FiberBundle.html Weisstein, Eric W. "Fiber bundle." From MathWorld--A Wolfram Web Resource. http://mathworld.wolfram.com/FiberBundle.html
p. 145: author
p.149: commons.wikimedia.org/wiki/File:William_Rowan_Hamilton_portrait_oval_combined.png public domain
p. 150: author
p.158: en.wikipedia.org/wiki/file:Sphere_rotation_qtl1.svg CC BY-SA 4.0 User:Quartl User:Masur, edited
S.160: Youtube: Visualizing quaternions (3blue1brown) screenshot, Ben Eater
p. 161: www.youtube.com/watch?v=pWOMDm6ejlw stereographic projection YouTube Васил Гергински
p.164: commons.wikimedia.org/wiki/File:Sphere_wireframe_10deg_6r.svg CC BY 3.0 User:Geek3
p. 165: commons.wikimedia.org/wiki/File:Torus_cycles.svg public domain
p. 165: commons.wikimedia.org/wiki/File:Torus-vill-point.svg CC BY-SA 4.0 User:Ag2gaeh
p. 167: commons.wikimedia.org/wiki/File:Young_Poincare.jpg public domain
p. 168: commons.wikimedia.org/wiki/File:Ricci_flow.png public domain
p. 170: www.youtube.com/watch?v=AKotMPGFJYk youtube screenshot Hopf Niles Johnson
p. 172: commons.wikimedia.org/wiki/File:Image_Tangent-plane.svg public domain
p. 180: commons.wikimedia.org/wiki/File:Stern-Gerlach_experiment_svg.svg CC BY-SA 4.0   Theresa Knott
p. 185: Möbius strip Joy Christian, arxiv.org/abs/1911.11578
p. 189: commons.wikimedia.org/wiki/File:World_line.svg CC BY-SA 3.0 User:K. Aainsqatsi User:Stib
p.202: commons.wikimedia.org/wiki/File:Portrett_av_Sophus_Lie.jpg public domain
p. 206: Author
Cover picture: Screenshot Visualizing quaternions, YouTube

END MARKS

1. A nice illustration can be found in the videos of Carl Sagan: Carl Sagan on Epicycles, Ptolemy, and Kepler (XXXSDESDEXXX).
2. Singh (2005), p.56
3. Rosenthal-Schneider (1988), p.24ff.
4. For example in M. J. Duff, L. B. Okun, G. Veneziano, arxiv.org/abs/physics/0110060.
5. Andre K.T.Assis; K.H.Wiederkehr; G.Wolfschmidt, *Weber's Planetary Model of the Atom*, Apeiron Montreal 2018, www.ifi.unicamp.br/~assis/Webers-Planeten-Modell-des-Atoms.pdf.
6. G. Lochak, *de Broglies initial concept of de Broglie waves*, in Diner (Ed.), The Wave-Particle Dualism, Springer Netherlands 1983, p. 1 ff.
7. G. Kirchhoff, *Philosophical Magazine* 13 (1857), pp. 393-412.
8. Feynman, *QED – The strange theory of Light and Matter* (1985), p.129.
9. E.g. Scholkmann et. al. (2017), iopscience.iop.org/article/10.1209/0295-5075/117/62002/meta
10. For more details see Unzicker, *Bankrupting Physics* (2013).
11. P.A.M. Dirac, *Nature* 139 (1937), p.323; Dirac, *Proc. Soc. London*, 165 (1938), p. 199 ff.
12. However, there are interesting observations pointing to a fractal distribution of galaxies in the Universe that question the definition of an average density. F. Sylos Labini, arXiv.org/abs/1103.5974 and arXiv.org/abs/1110.4041.
13. For more details see *Einstein's Lost Key*, chapters 4 and 5.
14. Schrödinger, *Annalen der Physik* 382 (1925), p.325-336.
15. Illustrative examples on this subject in Unzicker (2015), p. 76.
16. D. Sciama, *Monthly Notices of the Royal Astronomical Society*, Bd. 113, S. 34.
17. R. Dicke, *Rev.Mod.Phys* 29 (1957), pp. 363-376.
18. As stated in A. Unzicker, Ann. Phys. (Berlin) 18 (1), 57-70 (2009), arxiv.org/abs/07083518.
19. J. Broekaert, arXiv.org/abs/gr-qc/0405015; H. Dehnen et al. *Annalen der Physik* 461(1960), pp. 370-406; K. Krogh, arXiv.org/abs/astro-ph/9910325; M. Arminjon, arXiv.org/abs/gr-qc/0409092; H. E. Puthoff, arXiv.org/abs/9909037.

20 See an idea of my student Jan Preuss, arxiv.org/abs/1503.06763.
21 See also A. Unzicker, www.arXiv.org/abs/gr-qc/0702009.
22 *Proc. Roy. Soc.* London, 165 (1938), p. 199 ff.
23 R. Dicke, Gravitation without a principle of equivalence, *Rev.Mod.Phys* 29 (1957), p. 374.
24 A. Unzicker, *Dicke's momentous error*, vixra.org/abs/1510.0082.
25 This was first published in A. Unzicker, Annalen der Physik 18 (2009), pp. 53-70, see also Unzicker (2015), Chapter 10.
26 P.A.M. Dirac, *Nature*, Vol. 192, No. 4801, p. 441 (1961), followed by Dicke's reply.
27 Moreover, the effect is much less pronounced than earlier claims, T. Nielsen, A. Guffanti & S. Sarkar, www.nature.com/articles/srep35596.
28 See Unzicker (2015), chapter 12.
29 W. Finkelnburg, *Natural Sciences* 34 II (1947), 1947, p. 53ff.
30 With notable exception Dürr, H.-P., *New developments in high energy physics – the end of reductionism?* [German: Neuere Entwicklungen in der Hochenergiephysik - das Ende des Reduktionismus?] In: Self-Organization - The Emergence of Order in Nature and Society, (ed. A. Dress et. al. Munich 1986, pp. 15 - 34)
31 Mozkowski, A. (1921), loc. 2810.
32 Pohl, R., et al.(2010), *Nature*. 466 (7303) 213-216; W. Xiong et al. www.nature.com/articles/s41586-019-1721-2.
33 For a similar line of argument, see Unzicker (2013), pp. 255.
34 See however Williamson, J.G. et. al. Is the electron a photon with toroidal topology? *Annales de la Fondation Louis de Broglie*, Vol. 22, no.2, 133 (1997)
35 cf. G. Lochak (1983)
36 cf. Unzicker, www.arXiv.org/abs/gr-qc/0702009.
37 Kragh (2011), p. 177f.
38 see Unzicker (2012), p. 122ff.
39 A quote from A. Staruszkiewicz, Concepts of Physics 1 (2004):169
40 https://en.wikipedia.org/wiki/Aberration_(astronomy)
41 cf. Heisenberg, Physics and Beyond (1971).
42 D. Frauchinger and R. Renner, Quantum mechanics cannot describe consistently the use of itself,arxiv.org/pdf/1604.07422.pdf
43 cf. A. Unzicker, arXiv.org/abs/gr-qc/0011064.
44 Ehrenfest, *Zeitschrift für Physik* 78 (1932), pp. 555 - 559.
45 Maxwell (1873), volume 2, chapter 9, art. 618-619.

[46] Quaternions, Maxwell, Equations and Lorentz Transformations, M. Acevedo M., J. López-Bonilla and M. Sánchez-Meraz, Apeiron 12 (2005), S. 271-384; Doug Sweetser, https://www.science20.com/standup_physicist/blog/deriving_maxwell_source_equations_using_quaternions_25-82785
[47] Feyman, Richard (1999), The Pleasure of Finding Things Out, Perseus Publishing, pp.200-201.
[48] A. Gsponer and J.-P. Hurni, arxiv.org/abs/math-ph/0201058, p.8
[49] A very good reference is Wikipedia: Rotation formalisms in three dimensions.
[50] https://eater.net/quaternions/video/intro.
[51] The solid ball is three-dimensional, since its surface is S^2.
[52] In addition, there is a nice visualization YouTube: Dimensions Ep.7 Fibration I (Васил Гергински)
[53] A nice animation can be found on YouTube: Belt Trick (Jason Hise)
[54] The British mathematician Joy Christian considers another 7-dimensional manifold, see arxiv.org/abs/1806.02392.
[55] Pais, Inward Bound (Clarendon Press, 1986), p. 388.
[56] R. Penrose, *The Road to Reality* (2005), S. 619.
[57] J. Christian, arxiv.org/abs/1806.02392, arxiv.org/abs/1911.11578.
[58] J. Christian, arxiv.org/abs/1704.02876
[59] B. O'Sullivan, arxiv.org/abs/1601.02569;arxiv.org/abs/1611.02569.
[60] Einstein, Albert (1917). Sitzungsb. König. Preuss. Akad. 142–152.
[61] Cf. A. Unzicker, arxiv.org/abs/gr-qc/0011064.

www.ingramcontent.com/pod-product-compliance
Ingram Content Group UK Ltd.
Pitfield, Milton Keynes, MK11 3LW, UK
UKHW041352221225
9718UKWH00025B/172